Strengthening Disaster Risk Governance to Manage Disaster Risk

Strengthening Disaster Risk Governance to Manage Disaster Risk

Edited by

José Manuel Mendes
University of Coimbra, Centre for Social Studies, Faculty of Economics, Portugal

Gretchen Kalonji
Sichuan University - the Hong Kong Polytechnic Institute for Disaster Management and Reconstruction

Rohit Jigyasu
International Centre for the Study of the Preservation and Restoration of Cultural Property (ICCROM), Rome, Italy

Alice Chang-Richards
Department of Civil and Environmental Engineering
The University of Auckland, New Zealand

Elsevier
Radarweg 29, PO Box 211, 1000 AE Amsterdam, Netherlands
The Boulevard, Langford Lane, Kidlington, Oxford OX5 1GB, United Kingdom
50 Hampshire Street, 5th Floor, Cambridge, MA 02139, United States

Library of Congress Cataloging-in-Publication Data
A catalog record for this book is available from the Library of Congress

British Library Cataloguing-in-Publication Data
A catalogue record for this book is available from the British Library

ISBN: 978-0-12-818750-0

For information on all Elsevier publications
visit our website at https://www.elsevier.com/books-and-journals

Publisher: Candis Janco
Acquisitions Editor: Peter J Llewellyn
Editorial Project Manager: Sam W Young
Production Project Manager: Debasish Ghosh
Cover Designer: Christian J Bilbow

Typeset by SPi Global, India

Contents

Contributors

Numbers in parenthesis indicate the pages on which the authors' contributions begin.

Dilanthi Amaratunga (27), University of Huddersfield, Huddersfield, United Kingdom

Namrata Bhattacharya-Mis (59), University of Chester, Chester, United Kingdom

Laura Cevallos-Merki (107), Department of Geography, University of Zurich, Zürich, Switzerland

Graciela Fernández de Córdova (117), Department of Architecture, Pontifical Catholic University of Peru, Lima, Perú

Qingqing Feng (91), Department of Civil Engineering, The University of Hong Kong, Hong Kong, China

Karla Figueroa (77), Landscape and Urban Resilience Lab (PRULAB), Austral University of Chile, Valdivia, Chile

Richard Haigh (27), University of Huddersfield, Huddersfield, United Kingdom

Kinkini Hemachandra (27), University of Huddersfield, Huddersfield, United Kingdom

Rohit Jigyasu (21), International Centre for the Study of the Preservation and Restoration of Cultural Property (ICCROM), Rome, Italy

Jonas Joerin (107), Singapore-ETH Centre, Future Resilient Systems, Singapore, Singapore

Jessica Lamond (59), University of West of England, Bristol, United Kingdom

Mark R. Lapus (133), Agriculture Sustainability Initiatives for Nature, Inc., Bulacan, Philippines

Champika Liyanage (47), University of Central Lancashire, Lancashire, United Kingdom

S. Thomas Ng (91), Department of Civil Engineering, The University of Hong Kong, Hong Kong, China

Carolina Quintana (77), Landscape and Urban Resilience Lab (PRULAB), Austral University of Chile, Valdivia, Chile

Emmanuel Raju (69), Global Health Section & Copenhagen Centre for Disaster Research, University of Copenhagen, Copenhagen, Denmark; African Centre for Disaster Studies, North-West University, Potchefstroom, South Africa

Ortwin Renn (1), Institute for Advanced Sustainability Studies (IASS), Potsdam, Germany

Raymond S. Rodolfo (133), Ateneo de Manila University, Quezon City; Agriculture Sustainability Initiatives for Nature, Inc., Bulacan, Philippines

Sandra Santa-Cruz (117), GERDIS Research Group, Department of Engineering, Pontifical Catholic University of Peru, Lima, Perú

Juan N. Urteaga-Tirado (117), GERDIS Research Group, Department of Engineering, Pontifical Catholic University of Peru, Lima, Perú

Meghan Venable-Thomas (7), Enterprise Community Partners, Boston, MA, United States

Marta Vilela (117), Department of Architecture, Pontifical Catholic University of Peru, Lima, Perú

Paula Villagra (77), Institute of Environmental Science and Evolution; Landscape and Urban Resilience Lab (PRULAB), Austral University of Chile; Centre for Fire and Socioecosystem Resilience (FireSES), Valdivia, Chile

Felix Villalba-Romero (47), EAE Business School, Barcelona, Spain

Frank J. Xu (91), Department of Civil Engineering, The University of Hong Kong, Hong Kong, China

Carlos Zeballos-Velarde (99), Universidad Catolica de Santa Maria, Arequipa, Peru

Contributors

Introduction

José Manuel Mendes[a], Gretchen Kalonji[b], Rohit Jigyasu[c], and Alice Chang-Richards[d]

[a]*University of Coimbra, Centre for Social Studies, Faculty of Economics, Portugal,* [b]*Sichuan University - the Hong Kong Polytechnic Institute for Disaster Management and Reconstruction,* [c]*International Centre for the Study of the Preservation and Restoration of Cultural Property (ICCROM), Rome, Italy,* [d]*Department of Civil and Environmental Engineering, The University of Auckland, New Zealand*

Disaster risk governance is crucial for the implementation of effective disaster risk reduction (DRR) strategies. Articulation of institutional arrangements and public policies needs to take place in a multi-stakeholder context which incorporates community participation and full involvement of both private and public sectors at different levels (Clark-Ginsberg, 2020). However, as Trias and Cook (2019) concluded, the challenge is for the national government to arrange for the convergence of institutional capacities to empower local governments, and this is particularly true for southeast Asian countries.

The centrality of risk governance may not be fully recognized in the discussion papers, reports and policy recommendations of many organizations, including the Global Risks Report 2020 of the World Economic Forum (World Economic Forum, 2020). Although that report addresses risks in an unsettled global landscape, striving for economic stability and social cohesion, it does not recognize the importance of risk governance as an integrative approach to institutional coordination for disaster management.

This book aims to fill this gap by suggesting a new approach which contributes to the mainstreaming of risk governance in DRR strategies. The main focus is on the relevance of disaster risk governance in managing complex risks while also governing the everyday aspects of life and encouraging socio-economic development (UNDRR, 2019a,b).

The book presents a selection of contributions presented at the 8th International Conference on Building Resilience in Lisbon in 2018. It is structured according to the second principle of the Sendai Framework for Disaster Risk Reduction, 2015–2030, which is focused on *Strengthening Disaster Risk Governance to Manage Disaster Risk.* Therefore, the book includes discussion of risk and resilience from both a theoretical perspective, as well as innovative tools and good practices in reducing risk and building resilience. Combining the applications of social, financial, technological, urban planning, engineering and nature-based approaches, this volume addresses rising global priorities and focuses on building an improved understanding of risk governance trends, practices and initiatives.

The book is composed of 13 chapters. In the first chapter, Ortwin Renn proposes the concept of inclusive resilience as a new approach to risk governance. Renn works from the definition of resilience as "the capability of a socio-technical system to cope with events that are uncertain and ambiguous." He further analyzes management styles according to three types of resilience: adaptive, coping and participative. For Renn, risk governance depends on the ability to resolve complexity, characterize uncertainty and handle ambiguity. The chapter details the characteristics of four different ways of processing risk information according to risk complexity, scientific uncertainty and socio-political ambiguity, specifically: instrumental processing, epistemic processing, reflective processing and participative processing. Renn highlights the need to understand both objective and subjective natures of resilience before arriving at an adequate definition of risk governance.

The second chapter, authored by Meghan Venable-Thomas, presents cases of promoting community resilience through art and culture to improve health and well-being. To work with marginalized groups and communities, the concept of cultural resilience is essential to be included in people-centered interventions for people to cope with oppression, violence and adverse socioeconomic conditions. Venable-Thomas presents an innovative approach of interpretative epistemological analysis in five community-based organizations in the United States chosen to implement and manage climate change and resilience through creative placemaking. The important conclusion from Venable-Thomas' comparative analysis is that holistic community resilience must incorporate economic resilience and social capital, rather than just climate and cultural resilience. Also, she concludes by stressing the importance of attending to the structural and long-term tendencies in community disinvestment.

Rohit Jigyasu, in Chapter 3, addresses the importance of mainstreaming cultural heritage in disaster risk governance. The main argument is that the increased vulnerability of cultural heritage demands greater collaboration between agents in risk management, cultural heritage conservation and development, recurring also to traditional governance systems. Community engagement is crucial for achieving long-term sustainability and for governance based on the principles of collaboration, transparency, accountability and social justice. Rohit Jigyasu stresses the specificity of a governance approach that encompasses a broader perspective and mobilizes a great diversity of actors, mainly in matters concerning disaster risk governance for cultural heritage. The chapter ends with a discussion of the fundamental prerequisites for good governance and for mainstreaming cultural heritage in disaster risk governance.

In Chapter 4, Kinkini Hemachandra, Richard Haigh and Dilanthi Amaratunga analyze the role of higher education institutions in fostering multi-hazard early warning (MHEW) for coastal zones in Asia. The analysis is based on a thorough literature review emphasizing the crucial role of early warning for effective disaster preparedness, and planning for response and recovery. The authors identify 16 enablers of multi-hazard early warning for coastal resilience and aggregate them in three major categories, namely: policy, legislative and institutional arrangements; social and cultural considerations, and; technological arrangements. The first category directly addresses the main concern of the book related to risk governance. The strategic roles of higher education institutions in MHEW are discussed and three types of enablers are identified for these institutions: awareness and education; advocacy and evidence-based policy making, and; conducting research. The chapter ends with an analysis of the important role of regional cooperation for effective coastal resilience in Asia.

Felix Villalba-Romero and Champika Liyanage present and discuss in Chapter 5 the increased relevance of Disaster Risk Financing (DRF) to tackle disasters induced by climate change. The chapter relies on a literature review and addresses two research questions: what is DRF and what financial instruments can be used to implement DRF strategies? The authors conclude that the main objective of DRF is to achieve financial resilience by establishing financial strategies and appropriate financial instruments targeted to certain disasters. Also, the role of governments and private sectors in DRF and the challenges they face are analyzed. Villalba-Romero and Liyanage present five types of financial instruments: ex-post financing; ex-ante financing; traditional instruments; innovative risk financing mechanisms; and, finally, they propose an integrated approach for private and public sectors in DRF.

Chapter 6 by Jessica Lamond and Namrata Bhattacharya-Mis analyzes flood resilience in households and businesses in England from the property owners' perspective. The chapter draws on a survey across several locations that suffer from frequent floods in England. Both for households and businesses the percentage of flood insurance uptake is high although differences arise due to terms of the contracts and exclusions. These differences depend on the type of property and past flood experience. The authors also discuss the emergence of Flood Reinsurance (Flood RE) and its impact on small and medium-sized enterprises (SMEs). In contrast, homeowners tend to rely more on flood insurance while businesses are more proactive in taking measures to limit damage. The authors highlight the importance of adequate advice on risk assessment and management to mitigate flood impacts.

Emmanuel Raju in Chapter 7 compares the coordination processes between disaster recovery and response. Disaster recovery coordination emerges as an under-researched area. Raju uses the 2004 tsunami in Tamil Nadu, India as a case study to further his qualitative analysis. The main factors that influence different outcomes of disaster response and recovery coordination efforts include stakeholders' mandates, level of engagement, type of information received and objective coordination. Raju suggests that disaster recovery coordination should be linked with a broader discussion on sustainable development.

In Chapter 8, Paula Villagra Carolina Quintana and Karla Figueroa study the role of regulatory frameworks in building community resilience. Villagra, Quintana and Figueroa compare the governance approach with the governability approach, using a case study of the coastal zones of Southern Chile. The author utilize a variety of methodologies, including content analysis, spatial analysis, semantic networks and resilience capacity maps. The results show that there is a positive correlation between regulatory orientation and resilience. In the studied areas, governance is more present than governability, and therefore territorial resilience is more dependent on formal institutions than community knowledge or expertise.

The objective of Qingqing Feng, S. Thomas Ng and Frank J. Xu in Chapter 9 is to provide a theoretical basis for the harmonization of policies toward regional resilience in Guandong-Hong Kong-Macau Greater Bay Area in China. They identify policy instruments for regional resilience and evaluate them within a framework for adaptation and resilience policy analysis (FARP). Methodologically, hazards were identified through an online questionnaire and confirmed from interviews with selected participants. The authors emphasize the need to increase cross-city policy-making cohesiveness. They conclude by pointing out the main aspects for future research to address improvement of regional resilience in the studied areas.

Carlos Zeballos-Velarde proposes in Chapter 10 an integrated risk management model using participatory geographic information system (PGIS). The model is applied in the three phases of disaster risk management, namely: prospective, responsive and corrective phases, using the case study of city of Arequipa in Peru. Zeballos-Velarde describes the components of the different phases. The prospective management phase consisted of data compilation and analysis in GIS and the development of a district risk plan. The reactive phase resulted in the development of a geo application. The corrective management phase included a participatory workshop, the definition of critical sectors, an innovative drone survey and subsequent modeling, vulnerability calculation and model validation. Zeballos-Velarde concludes by highlighting the importance of community participation in all phases of disaster risk management.

In Chapter 11 Laura Cevallos-Merki and Jonas Joerin analyze the role of social capital in disaster recovery. Their case study is situated in the aftermath of the 2016 earthquake in Ecuador. By using quantitative and qualitative methodologies, they conclude that vertical social capital, which is trust and access to authority representatives, is positively correlated with post-disaster recovery satisfaction. On the other hand, social capital at the community and individual levels seem to have no correlation with level of recovery satisfaction. Collaboration in the studied areas was investigated along the lines of family networks and networks among friends and neighbors. These results led the authors to recommend strengthening of community trust and organization by creating and consolidating community-based organizations with the help and support from local councils.

Juan N. Urteaga-Tirado, Sandra Santa-Cruz, Graciela Fernández de Córdova and Marta Vilela advance in Chapter 12 an ex post methodology to analyze the influence of hazard mitigation plans in fostering local resilience. Their empirical context is coastal localities in Peru affected by El Niño heavy rainfalls in 2017. Their study includes an analysis of hazard mitigation plans at two levels. The first level is based on an extensive analysis structured along the application of four indices that are correlated with resilience variables. The second level is a comprehensive analysis based on reviews and interviews with selected actors from a sub-sample of the localities under study. The authors conclude that reduction of damages should be included in action plans and identification of the responsible persons at different levels is paramount for post-disaster reconstruction process.

The last chapter in this book is authored by Raymond Rodolfo and Mark Lapus. Their focus is on business continuity initiatives in DRR, using an example of oyster farmers in the Philippines. Their results are based on two semi-structured focus group discussions. They emphasized the importance of collaboration between community members, business sectors and government agencies to develop more sound and sustainable business practices as part of community resilience and DRR initiatives.

References

Clark-Ginsberg, A., 2020. Disaster risk reduction is not "everyone's business": evidence from three countries. Int. J. Disaster Risk Reduct. https://doi.org/10.1016/j.ijdrr.2019.101375.

Trias, A., Cook, A., 2019. Recalibrating disaster governance in ASEAN: implications of the 2018 Central Sulawesi earthquake and tsunami. Research report, S. Rajaratnam School of International Studies. Retrieved from: https://www.jstor.org/stable/resrep20024?refreqid=excelsior%3Acbef07252179197ba26b3f39eaa6db0a.

United Nations Office for Disaster Risk Reduction (UNDRR), 2019a. Annual report. UNDRR, Geneva.

United Nations Office for Disaster Risk Reduction (UNDRR), 2019b. Global Assessment Report on Disaster Risk Reduction (GAR). UNDRR, Geneva.

World Economic Forum, 2020. Global Risks Report 2020. Insight report, fifteenth ed. World Economic Forum, Marsh & McLennan and Zurich Insurance Group, Geneva. Retrieved from: http://www3.weforum.org/docs/WEF_Global_Risk_Report_2020.pdf.

Chapter 1

Inclusive resilience: A new approach to risk governance

Ortwin Renn
Institute for Advanced Sustainability Studies (IASS), Potsdam, Germany

1 Introduction

The concept of resilience has been used in may disciplines for different notions of being able to respond adequately when the system is under stress. It has been widely applied in ecological research and denotes the resistance of natural ecosystems to cope with stressors (Holling, 1973). Resilience is focused on the ability and capacity of systems to resist shocks and to have the capability to deal and recover from threatening events (Rose, 2007; Jackson and Ferris, 2017). This idea of resistance and recovery can also be applied to social systems (Review in Norris et al., 2008; Adger, 2000). The main emphasis here is on organizational learning and institutional preparedness to cope with stress and disaster. The US Department of Homeland Security (DHS) uses this definition: "Resilience is the ability of systems, infrastructures, government, business, and citizenry to resist, absorb, and recover from or adapt to an adverse occurrence that may cause harm, destruction, or loss [that is] of national significance" (cited after Longstaff et al., 2010, p. 19). Hutter (2011) added to this analysis the ability of systems to respond flexibly and effectively when a system is under high stress from unexpected crisis. Pulling from an interdisciplinary body of theoretical and policy-oriented literature, Longstaff et al. (2010) regard resilience as a function of resource robustness and adaptive capacity.

The governance framework suggested by the International Risk Governance Council (IRGC, 2017) depicts resilience as a normative goal for risk management systems to deal with highly uncertain events or processes (surprises). It is seen as a property of risk-absorbing systems to withstand stress (objective resilience) but also the confidence of risk management actors to be able to master crisis situations (subjective resilience).

In this chapter I explain the connection between inclusiveness of risk governance based on multiple stakeholder involvement, and the need to enhance resilience, understood here as the capability of a socio-technical system to cope with events that are uncertain and ambiguous (Renn and Klinke, 2016). This approach has been inspired by Lorenz (2010), who distinguishes adaptive, coping and participative aspects of resilience. I will use this classification to discern between three management styles which correspond to these three aspects of resilience. I have called them: risk-informed (corresponding to adaptive capability); precaution-based (corresponding to coping capability) and discourse-based (corresponding to participative capability).

2 Complexity, uncertainty and ambiguity in risk governance

Understanding and managing risks is confronted with three major challenges: complexity, uncertainty and ambiguity (Klinke and Renn, 2019; Rosa et al., 2014, p. 130ff). Complexity refers to the difficulty of identifying and quantifying causal links between a multitude of potential candidates and specific adverse effects. Uncertainty denotes the inability to provide accurate and precise quantitative assessments between a causing agent and an effect. Finally, ambiguity denotes either the variability of (legitimate) interpretations based on identical observations or data assessments or the variability of normative implications for risk evaluation (judgment on tolerability or acceptability of a given risk).

In a case where scientific complexity is high and uncertainty and ambiguity are low, the challenge is to invite experts to deliberate with risk managers to understand complexity. Understanding the risks of oil platforms may be a good example of this. Although the technology is highly complex and many interacting devices lead to multiple accident scenarios most possible pathways to a major accident can be modeled well in advance. The major challenge is to determine the limit to which one is willing to invest in resilience.

Strengthening Disaster Risk Governance to Manage Disaster Risk. https://doi.org/10.1016/B978-0-12-818750-0.00001-5

The second route concerns risk problems that are characterized by high uncertainty but low ambiguity. Expanded knowledge acquisition may help to reduce uncertainty. If, however, uncertainty cannot be reduced (or only reduced in the long run) by additional knowledge, a "precaution-based risk management" is required. Precaution-based risk management explores a variety of options: containment, diversification, monitoring, and substitution. The focal point here is to find an adequate and fair balance between over cautiousness and insufficient caution. This argues for a reflective process involving stakeholders to ponder concerns, economic budgeting, and social evaluations.

For risk problems that are highly ambiguous (regardless of whether they are low or high on uncertainty and complexity), route 3 recommends a "discourse-based management." Discourse management requires a participatory process involving stakeholders, especially the affected public. The aim of such a process is to produce a collective understanding among all stakeholders and the affected public about how to interpret the situation and how to design procedures for collectively justifying binding decisions on acceptability and tolerability that are considered legitimate. In such situations, the task of risk managers is to create a condition where those who believe that the risk is worth taking and those who believe otherwise are willing to respect each others' views and to construct and create strategies acceptable to the various stakeholders and interests.

In essence: The effectiveness and legitimacy of the risk governance process depends on the capability of management agencies to resolve complexity, characterize uncertainty and handle ambiguity by means of communication and deliberation.

3 Instrumental processing involving governmental actors

Dealing with linear risk issues, which are associated with low scores for complexity, scientific uncertainty and socio-political ambiguity, requires hardly any changes to conventional public policy-making. The data and information regarding such linear (routine) risk problems are provided by statistical analysis; law or statutory requirements determine the general and specific objectives; and the role of public policy is to ensure that all necessary safety and control measures are implemented and enforced (Klinke and Renn, 2012). Traditional cost-benefit analyses including effectiveness and efficiency criteria are the instruments of political choice for finding the right balance between under- and over-regulation of risk-related activities and goods. In addition, monitoring the area is important to help prevent unexpected consequences. For this reason, linear risk issues can well be handled by departmental and agency staff and enforcement personnel of state-run governance institutions. The aim is to find the most cost-effective method for a desired regulation level. If necessary, stakeholders may be included in the deliberations as they have information and know-how that may help to make the measures more efficient.

4 Epistemic processing involving experts

Resolving complex risk problems requires dialogue and deliberation among experts. The main goal is to scan and review existing knowledge about the causal connections between an agent and potential consequences, to characterize the uncertainty of this relationship and to explore the evidence that supports these inferences. Involving members of various epistemic communities which demonstrate expertise and competence is the most promising step for producing more reliable and valid judgments about the complex nature of a given risk. Epistemic discourse is the instrument for discussing the conclusiveness and validity of cause-effect chains relying on available probative facts, uncertain knowledge and experience that can be tested for empirical traceability and consistency. The objective of such a deliberation is to find the most cogent description and explanation of the phenomenological complexity in question as well as a clarification of dissenting views (for example, by addressing the question which environmental and socio-economic impacts are to be expected in which areas and in what time frame). The deliberation among experts might generate a profile of the complexity of the given risk issue on selected inter-subjectively chosen criteria. The deliberation may also reveal that there is more uncertainty and ambiguity hidden in the case than the initial appraisers had anticipated (Birkmann, 2011). It is advisable to include natural as well as social scientists in the epistemic discourse so that potential problems with risk perception and risk frames can be anticipated. Controversies would then be less of a surprise than is currently the case. Such epistemic discourse is meant to lead to adaptive management procedures that monitor the state of knowledge and proficiency in the field and adjust management responses according to the various levels of knowledge available at each time period (Wiering and Arts, 2006; Henwood and Pidgeon, 2016).

5 Reflective processing involving stakeholders

Characterizing and evaluating risks as well as developing and selecting appropriate management options for risk reduction and control in situations of high uncertainty poses particular challenges. How can risk managers characterize and evaluate the severity of a risk problem when the potential damage and its probability are unknown or highly uncertain? Scientific input is, therefore, only the first step in a series of steps constituting a more sophisticated evaluation process. It is crucial to compile the relevant data and information about the different types of uncertainties to inform the process of risk characterization. The outcome of the risk characterization process provides the foundation for a broader deliberative arena, in which not only policy makers and scientists, but also directly affected stakeholders and public interest groups ought to be involved in order to discuss and ponder the "right" balances and trade-offs between over- and under-protection (Renn and Schweizer, 2009). This reflective involvement of stakeholders and interest groups pursues the purpose of finding a consensus on the extra margin of safety that potential victims would be willing to tolerate and potential beneficiaries of the risk would be willing to invest in to avoid potentially critical and catastrophic consequences. If too much precaution is applied, innovations may be impeded or even eliminated; if too little precaution is applied, society may experience the occurrence of undesired consequences. The crucial question here is how much uncertainty and ignorance the main stakeholders and public interest groups are willing to accept or tolerate in exchange for some potential benefit.

This issue has direct implications for resilience. As this concept reflects the confidence of all actors to deal with even uncertain outcomes, it provides a mental guideline for the negotiations between beneficiaries and potential victims of risks (IRGC, 2017). Furthermore, it includes a discourse about coping capacity and compensation schemes if the worst were to happen. The boundary between subjective and objective resilience is, however, fuzzy under the condition of effect uncertainty (Brown and Kulig, 1996/97; Norris et al., 2008). In cases of known risks past experience can demonstrate whether the degree of self-confidence was accurate and justified. Over long time spans one would expect an emerging congruence between objective and subjective resilience (learning by trial and error). However, for extremely rare events or highly uncertain outcomes, one necessarily relies on models of anticipation and expectations that will widely vary among different stakeholder groups, in particular those who benefit and those who will bear the risks. Furthermore, there will be lots of debates about the potential distribution of effects over time and space. The degree of coping capacity that is regarded as sufficient or justified for approving a new risk agent or a disaster management plan to become enacted depends therefore on a discourse between the directly affected groups of the population. Such a reflective involvement of policy makers, scientists, stakeholders and public interest groups can be accomplished through a spectrum of different procedures such as negotiated rule-making, mediation, round-table or open forums, advisory committees, and others (see Beierle and Cayford, 2002; Benn et al., 2009; Renn, 2015).

6 Participative processing involving the wider public

If risk problems are associated with high ambiguity, it is not enough to demonstrate that risk regulation addresses the public concerns of those directly affected by the impacts of the risk source. In these cases, the process of evaluation and management needs to be open to public input and new forms of deliberation. This corresponds with the participative aspect of resilience (Lorenz, 2010). Such discursive activities should start with revisiting the question of proper framing. Is the issue really a risk problem or is it an issue of lifestyle or future vision? Often the benefits are contested as well as the risks. The debate about "designer babies" may illustrate the point that observers may be concerned not only about the social risks of intervening in the genetic code of humans but also about the acceptability of the desired goal to improve the performance of individuals (Hudson, 2006). Thus the controversy is often much broader than dealing with the direct risks only. The aim here is to find an overlapping consensus on the dimensions of ambiguity that need to be addressed in comparing risks and benefits, and balancing pros and cons. High ambiguity would require the most inclusive strategy for involvement because not only directly affected groups but also those indirectly affected should have an opportunity to contribute to this debate.

Resolving ambiguities in risk debates necessitates the participatory involvement of the public to openly discuss competing arguments, beliefs and values. Participatory involvement offers opportunities to resolve conflicting expectations through a process of identifying overarching common values, and to define options that will allow a desirable lifestyle without compromising the vision of others. Critical to success here is the establishment of equitable and just distribution rules when it comes to common resources and a common understanding of the scope, size and range of the problem, as well as the options for dealing with the problem (Renn and Schweizer, 2009). Unless there is some agreement on the boundaries of what is included, there is hardly any chance for a common solution. Such a common agreement will touch upon the coping capacity of systems to deal with different frames of risks and not only with the physical impacts of risks. There are various social constructions of resilience that the participants associate with the management options. The set of possible procedures for involving the public includes citizen panels or juries, citizen forums, consensus conferences, public advisory committees and similar approaches (see Rowe and Frewer, 2000; Beierle and Cayford, 2002; Renn, 2008, p. 284ff; Wong, 2018; Radtke et al., 2018).

7 Conclusions

The goal of this chapter has been to illustrate the significance of resilience for risk governance, including all stages from pre-assessment to management and communication. For this purpose, the resilience concept by Lorenz was applied to link risk governance strategies with the three major aspects of resilience: adaptive management capacity, coping capacity, and participative capacity. The three risk characteristics—complexity, uncertainty and ambiguity—were linked to these three aspects of resilience. Furthermore, the three aspects were used to develop four major risk management and discourse strategies; beginning with simple risk management in which none of these characteristics and capacity requirements were involved to discourse-based management in which all three characteristics and capacity requirements were combined.

Whereas the analysis of simple and—to some degree—complex problems is better served by relying on the physical understanding of experienced resilience, uncertain and ambiguous problems demand the integration of social constructions and mental models of resilience, operationalized as confidence in one's coping capacity, for both understanding and managing these problems. The distinction of risks according to risk characteristics not only highlights deficits in our knowledge concerning a risk issue, but also points the way forward for the selection of the appropriate management options. Thus, the risk governance framework attributes an important function to public and stakeholder participation, as well as risk communication, in the risk governance process. The framework suggests efficient and adequate public or stakeholder participation procedures. The concerns of stakeholders and/or the public are integrated in the risk appraisal phase via concern assessment. Furthermore, stakeholder and public participation are an established part of risk management. The optimum participation method depends on the characteristics of the risk issue. In this respect, the three aspects of resilience are gradually included into the various discourses. The need for finding an agreement on what constitutes an adaptive, coping and participative response to ensuring resilience underlines the necessity to understand and comprehend the objective and subjective nature of resilience.

References

Adger, W.N., 2000. Social and ecological resilience: are they related? Prog. Hum. Geogr. 24 (3), 347–364.

Beierle, T.C., Cayford, J., 2002. Democracy in Practice. Public Participation in Environmental Decisions. Resources for the Future, Washington, DC.

Benn, S., Dunphy, D., Martin, A., 2009. Governance of environmental risk: new approaches to managing stakeholder involvement. J. Environ. Manag. 9 (4), 1567–1579.

Birkmann, J., 2011. First- and second-order adaptation to natural hazards and extreme events in the context of climate change. Nat. Hazards 58 (2), 811–840.

Brown, D., Kulig, J., 1996/97. The concept of resiliency: theoretical lessons from community research. Health Can. Soc. 4, 29–52.

Henwood, K.L., Pidgeon, N., 2016. Interpretive environmental risk research: affect, discourses and change. In: Crichton, J., Candlin, C.N., Firkins, A.S. (Eds.), Communicating Risk. Communicating in Professions and Organizations. Palgrave Macmillan, London, pp. 155–170.

Holling, C.S., 1973. Resilience and stability of ecological systems. Annu. Rev. Ecol. Syst. 4 (1), 1–23.

Hudson, K.L., 2006. Preimplantation diagnosis: public policy and public attitudes. Fertil. Steril. 58 (6), 1638–1645.

Hutter, G., 2011. Organizing social resilience in the context of natural hazards: a research note. In: Hutter, G., Kuhlicke, C., Glade, T., Felgentreff, C. (Eds.), Natural Hazards, pp. 1–14. Special Volume: Resilience in Hazards Research and Planning—A Promising Concept, (11 January 2011).

IRGC (International Risk Governance Council), 2017. Risk Governance: Towards an Integrative Approach. White Paper, IRGC, Geneva.

Jackson, S., Ferris, L.J., 2017. Designing resilient systems. In: Linkow, I., Palma-Oliveira, J.M. (Eds.), Resilience and Risk. Springer, Heidelberg and New York, NY, pp. 121–144.

Klinke, A., Renn, O., 2012. Adaptive and integrative governance on risk and uncertainty. J. Risk Res. 1 (1), 3–20.

Klinke, A., Renn, O., 2019. The coming age of risk governance. Risk Anal. https://doi.org/10.1111/risa.13383.

Longstaff, P.H., Armstrong, N.J., Perrin, K., Parker, W.M., Hidek, M.A., 2010. Building resilient communities: a preliminary framework for assessment. Homeland Secur. Aff. VI (3), 1–23.

Lorenz, D.F., 2010. The diversity of resilience: contributions from a social science perspective. In: Hutter, G., Kuhlicke, C., Glade, T., Felgentreff, C. (Eds.), Natural Hazards, pp. 1–18. Special Volume: Resilience in Hazards Research and Planning—A Promising Concept? (November 23, 2010).

Norris, F.H., Stevens, S.P., Pfefferbaum, B., Wyche, K.E., Pfefferbaum, R.L., 2008. Community resilience as a metaphor, theory, set of capabilities, and strategy for disaster readiness. Am. J. Community Psychol. 41, 127–150.

Radtke, J., Holstenkamp, L., Barnes, J., Renn, O., 2018. Concepts, formats, and methods of participation: theory and practice. In: Holstenkamp, L., Radtke, J. (Eds.), Handbuch Energiewende und Partizipation. Springer VS, Wiesbaden, pp. 21–42.

Renn, O., 2008. Risk Governance. Coping with Uncertainty in a Complex World. Earthscan, London.

Renn, O., 2015. Stakeholder and public involvement in risk governance. Int. J. Disaster Risk Sci. 6 (1), 8–20.

Renn, O., Klinke, A., 2016. Complexity, uncertainty and ambiguity in inclusive risk governance. In: Andersen, T.J. (Ed.), The Routledge Companion to Strategic Risk Management. Routledge, Milton Park and New York, NY, pp. 13–30.

Renn, O., Schweizer, P., 2009. Inclusive risk governance: concepts and application to environmental policy making. Environ. Policy Gov. 19, 174–185.

Rosa, E.A., Renn, O., McCright, A.M., 2014. The Risk Society Revisited. Social Theory and Governance. Temple University Press, Philadelphia, PA.

Rose, A., 2007. Economic resilience to natural and man-made disasters: multidisciplinary origins and contextual dimensions. Environ. Hazards 7, 383–398.

Rowe, G., Frewer, L.J., 2000. Public participation methods: a framework for evaluation. Sci. Technol. Hum. Values 25 (1), 3–29.

Wiering, M.A., Arts, B.J.M., 2006. Discursive shifts in dutch river management: "deep" institutional change or adaptation strategy? Hydrobiologia 565 (1), 327–338.

Wong, C.M.L., 2018. From risk management to risk governance. In: Energy, Risk and Governance. The Case of Nuclear Energy in India. Palgrave Macmillan, Cham, pp. 199–248.

Chapter 2

Can cultural resilience be a tool for strengthening community?

Meghan Venable-Thomas
Enterprise Community Partners, Boston, MA, United States

1 Introduction

As a public health leader, I believe health is a foundational social benefit and seek to integrate this principle across disciplines and practices that support people's ability to live prosperous lives. Essential to making this reality is contributing to the field of community resilience. The Climate and Cultural Resilience (C&CR) program with *Enterprise Community Partners* attempted to approach resilience development through a diverse approach, using art and culture to create innovative community development practices that actively promote health and well-being among the most underserved and enhance climate practices to build healthier and more resilient communities. Additionally, this program sought to bring together sustainability and art practitioners to engage in new kinds of partnerships that foster the growth of more equitable and resilient communities.

In recent years, the "range of social and physical stresses that people and communities experience has multiplied" (Acosta et al., 2018). Additionally, acute events, combined with these stresses, affect many communities consistently over time. The scholarship and practice of ways to promote resilience for individuals and communities continues to grow. As the scope grows, so does the number of differing perspectives about the most important factors contributing to individual and community resilience (Acosta et al., 2018). However, if we can better understand what resilience means for the communities we serve and what solutions they are already employing to be resilient, we may be able to create and support more effective solutions. This project is an opportunity to explore one approach to fostering community resilience for expansion of such strategies in the future benefiting communities and stakeholder groups. The C&CR Program uses creative placemaking as a strategy for supporting community resilience. As shown in Fig. 2.1 *Enterprise* believes that investing in cultural resilience simultaneously with climate resilience will improve overall community resilience. The component of cultural resilience offers a more nuanced definition of community resilience that places "capacity to maintain and develop cultural identity and critical cultural knowledge and practices" (Clauss-Ehlers, 2010) as an equal contributor with climate to a community's overall resilience.

1.1 Community resilience

Despite nuances in the definition of community resilience, the underlying premise involves the idea of a group being able to recover from a significant event or prolonged stressor (Yosso, 2005). Grounding community resilience from a population health perspective recognizes that resilience is a process for supporting a broad range of positive physical and mental health outcomes related to socio-environmental exposures and that individual risk of illness or disease cannot be considered in isolation from the disease risk of the population to which it belongs (Berkman et al., 2015). Health inequities are prominent in urban centers across the United States and often, because of historical inequality, are related to race and income. Although race is a social construct, disinvestments, discrimination, and social devaluing committed over time, based on race, have created population-level health disparities in marginalized communities. This highlights the importance of understanding the social determinants of health in order to begin to overcome present community-based health inequities and disparities and support community resilience efforts.

1.2 Cultural resilience

Enterprise believes that cultural resilience is foundational to community resilience in supporting the people-based component of resilience. Shared cultural identities help people connect and empathize with others' experiences. Understanding culture resilience as a tool for connecting particularly marginalized groups to one another and the environment they exist

Strengthening Disaster Risk Governance to Manage Disaster Risk. https://doi.org/10.1016/B978-0-12-818750-0.00002-7

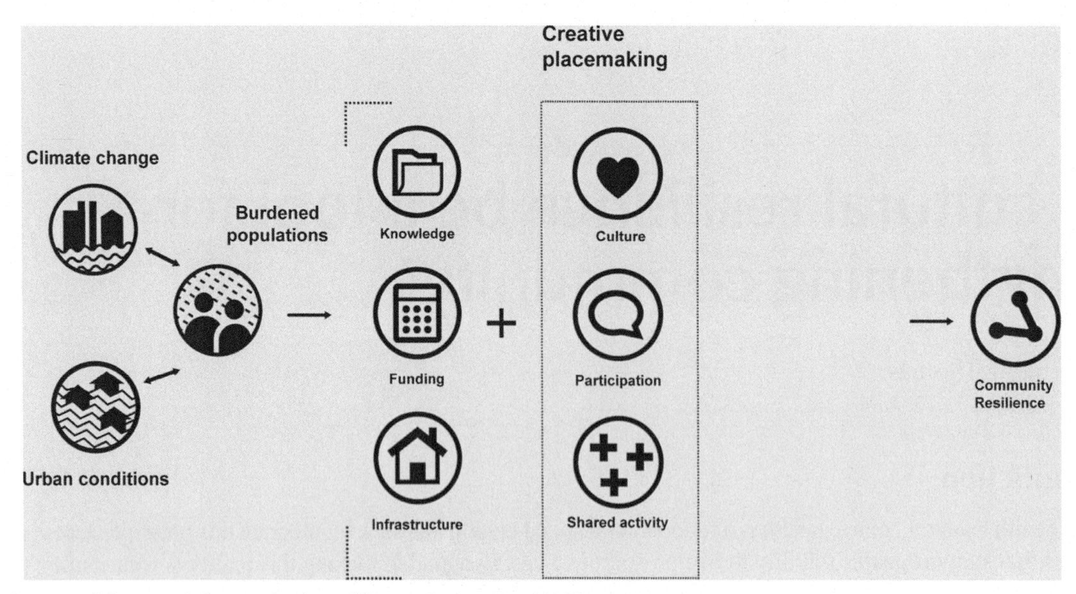

FIG. 2.1 Conceptual framework for community resilience. *(Source: Venable-Thomas.)*

within recognizes its ability as a community resilience component to uniquely support climate resilience challenges. The cultural resilience concept has been examined in several studies of groups responding to oppression, violence, and adverse socioeconomic conditions in countries around the world. Researchers cite the term cultural resilience as a mechanism leveraged by First Nations communities to promote protective mechanisms and behaviors in community youth by maintaining and reviving their cultural heritage (Lalonde, 2005). As we begin to think about why culture matters to resilience and community development efforts, it is important to know that discrimination, another social determinant of health, and the oppression of culture through structural and interpersonal prejudice has driven many of the health inequities we see today. Thus, incorporating ways to support and uplift cultural identity and social connectedness of underrepresented groups is an often forgotten but very important component for connection building and, in turn, community resilience.

The most well-established literature on social connection is in the realm of health and well-being (Putnam, 2000). Social capital has been linked to improved mental health (Berkman and Kawachi, 2001), decreased mortality (Kawachi et al., 1997), increased adolescent well-being (Howard, 2003), to name a few. One of the most well-known social capital studies presented by Daniel Aldrich in *Building Resilience* found that the presence of strong social capital, both among people and among individuals and organizations, is a prerequisite for and a predictor of recovery. Additionally, he claimed that social capital might be even more important to resilience than both the degree of infrastructure damage and the amount of aid received by an area (Aldrich, 2012). Building social cohesion and capital supports health resilience and thus is a critical element of not only building preparedness and facilitating recovery but also maintaining community resilience.

2 Objectives

The Climate and Cultural Resilience (C&CR) Program—which funded five community-based organizations across the United States to use creative placemaking toward community resilience outcomes—presented a component not seen in other creative placemaking programs. The component of cultural resilience offers a more nuanced definition of community resilience that places "capacity to maintain and develop cultural identity and critical cultural knowledge and practices" (Clauss-Ehlers, 2010) as an equal contributor as climate to advancing a community's overall resilience. *Enterprise* hired me in June 2016 to manage the pilot of the Climate & Cultural Resilience Program (C&CR). This study used the C&CR Program to interrogate the role of creative placemaking within community development projects. I approached this work by investigating the following:

- How climate resilience, cultural resilience and creative placemaking are understood among different stakeholders engaged in community development;

- The role of creative placemaking in advancing climate and cultural resilience; and
- The role that an intermediary might have in order to influence toward these strategies.

This study sought to answer these questions by exploring five different exemplars for using creative placemaking as a tool for building community resilience and analyzing their alignment with expressed community needs.

3 Methods

Through an interpretative epistemological approach I developed a qualitative case study, in which I simultaneously worked with theory and empirical material. I collected my findings by performing document analysis, focus groups, participant observation, and conducting semi-structured interviews with people participating in the C&CR Program. I chose focus groups as an effective tool for uncovering individual as well as group perceptions and opinion. I conducted 12 focus groups, which incorporated a total of 59 people. The following was the breakdown for each site: Community members and staff were in separate focus groups. The American Indian Community Housing Organization in Duluth, Minnesota: six community members, two staff members; WonderRoot in Atlanta, Georgia: 15 community members, two staff members; The Center for Neighborhood Technology in Chicago, Illinois: 15 community members, three staff members; Chinatown Community Development Center in San Francisco, California: six community members, two staff members; Coalfield Development Corporation in Wayne, West Virginia: six community members, two staff members. I followed up focus groups with 20 focused interviews based on previous observations and data collected during focus groups. The key interview questions focused on how the respondents understood and had experienced climate and cultural resilience and creative placemaking in their communities. Inclusion criteria for those in focus groups or interviewed was to be participants in the C&CR project directly as project leadership planning the project, community members impacted by the project, or designers implementing the projects. The only exclusion criterion was not participating in or not being impacted by the C&CR program projects. The participants were recommended from program leadership at each of the sites selected to receive the C&CR grants.

Since the aim of this study was to gain a deeper understanding of climate and cultural resilience and creative placemaking in the C&CR Program, qualitative research was the most suitable method. Following Stake (1995) and Yin (2003), I based my approach to the case study on a constructivist paradigm. This paradigm "recognizes the importance of the subjective human creation of meaning, but doesn't reject outright some notion of objectivity. Constructivism is built upon the premise of a social construction of reality" (Searle, 1995). One of the advantages of this approach is the close collaboration between the researcher and the participant, while enabling participants to tell their stories (Crabtree and Miller, 1992).

The *Enterprise* criteria for evaluating the projects of these organizations are as follows. A comprehensive overview of how each organization leveraged the criteria can be found in Fig. 2.2. Each of the grantees was evaluated on their ability to select and conduct one or more of those activities and what creative placemaking approaches they used to do so:

(1) Climate resilience advancing activities (Young, 2017):
  (a) Learn from residents about specific ways in which climate impacts affect them and the places they live;
  (b) Proactively address potential climate-related challenges for the particular place;
  (c) Plan to address specific climate resilience needs to be paired with cultural resilience efforts for added impact.
(2) Cultural resilience advancing activities:
  (a) Partner with artists or designers and with residents or community members to create an experience or product that reflects community identity;
  (b) Conduct community-engaged activities that focus on cultural expression of the people involved;
  (c) Use culturally competent practices to deliver the services of the organization and to gain stakeholder input.

4 Results

4.1 Grantee case study

Each of the grantee organizations presented resilience and cultural challenges within their communities. However, these challenges were in the context of the Request for Proposal (RFP) from *Enterprise* and were selected because they had experience conducting activities that advanced climate and cultural resilience. Therefore, these components are present in each of the grantee projects but the extent to which they are present varies across the grantee groups. Each of the grantee groups

Organization	Climate Resilience Strategy	Cultural Resilience Strategy	Creative Placemaking Approach	Outcomes
AICHO	Rooftop garden redesign & Capacity building for sustainability efforts: • Installation of a 12 kW photovolatic array • Installation of composting system • Installation of water capture system • Natural food production	Rooftop garden programming: • Native American public art • Creation of a C &CR speaker series on NA topic areas • Development of a C&CR Community Committee	• C&CR Community Committee to select art installation • Selected artist designed and installed mural • Conducted a community participatory design activity with residents and community to inform rooftop design	1B, 1C, 2A, 2B, 2C
WonderRoot & SouthFAce	Support the development of a local Climate Resiliency Plan to identify climate-related risks and opportunities, and help guide needs to build resilience: • Conduct a community-led arts-based needs assessment to identify climate facility needs	• Conduct a community-led arts-based needs assessment to identify cultural facility needs, and develop a cultural facility siting and operation plan within the Lee Street corridor	• Using creative placemaking platforms to elevate discussions about creating new sustainability and resiliency policies that are people-first • Design and launch of a public art/way-finding project that denotes community assets in the natural and built environment • Hire additional creative placemaking staff person Five key elements to the CP strategy that informed how they approached: 1. Cross sector partnership 2. Incorporating the social landscape of the area 3. Centered on a specific place- Lee St. Corridor 4. Publicly visible 5. Equitable community partnership	1A, 2A, 2B,
Chi-Go	Rain water infrastructure projects installed in 4 neighborhoods near metro stations: • Green Line (51 st St.) rain gardens on boxville community market • Blue Line (Holamn Square)-10 trees planted of a 7000 tree planting proj • Pink Line (California)-rain boxes at leaking overpass • Blue Line (Logan Square)-splashboxes for rain water runoff	Local cultural events positioned near metro stations w/climate infrastructure: • Green Line (51 st St.)-mural depicting the local black community • Blue Line (Holamn Square)- crafting of shovels by local artists from reclaimed weapons-highlighting the issues of community violence • Pink Line (California)- placement of public art gallery under overpass with local Mexican artist's work • Blue Line (Logan Square) creation of a mural by local youth depicting the cultural Puerto Rican assets of the community	Each area had a lead partner nonprofit that worked to: (1) Involve local artists; (2) Connect to local anchor institutions such as schools and houses of worship; and (3) Direct local implementation efforts with community residents and stakeholders. A community-led committee: • selected the local artists. • The artists then engaged in multiple community participatory sessions to identify and devlope public art installations, programming, and community gardens.	1B, 1C, 2A, 2B
Chinatown CDC	Develop a city plan and conceptual design for the redevelopment of Portsmouth Square Park that positions it to be a resilience hub for Chi natown residents	Develop a city plan and conceptual design for the redevelopment of Portsmouth Square Park that reflects the cultural of Chinatown residents	CCDC employed a community planning process for portsmouth Square that included: • four community meetings • one-on-one interview with stakeholders and • intercept surveys with users of the park.	1A, 1B, 1C, 2A, 2B, 2C
Coalfield	Addresses both climate and economic challenges by hiring out of work coal miners on projects that contribute to sustainable community development. Activates include: • Planting trees, • maintain and harvesting local produce, • installing a solar farm, and • repurposing wood from cut downtrees	Mentorship program by local artists for former coal miners in: • quilting woodworking, mountain music, foraging/canning, beekeeping, pottery, and glassblowing.	Every program incorporates a design process with a: • community design charette facilitated by LEED architeets • partnership with community groups to ensure broad outreach and involvement • All trainees are hired locally • Use local partnerships to select artist mentors who develop mentorship curriculum	1C, 2A, 2B

FIG. 2.2 C&CR program project overview. *(Source: Venable-Thomas.)*

has an organizational partner experienced in climate or cultural issues. Each lead organization is a community development corporation (CDC), which are non-profit, community-based organizations focused on revitalizing the areas in which they are located, typically low-income, underserved neighborhoods that have experienced significant disinvestment. While they are most commonly celebrated for developing affordable housing, they are usually involved in a range of initiatives critical to community health such as economic development, sanitation, streetscaping, neighborhood planning projects, education, and social service provision to neighborhood residents (CW, 2018). Through the C&CR project, *Enterprise* provided funding to support capacity building, as well as connecting them with resources, distributing best practices, publicizing successes, and connecting to provider networks.

This section takes a deep dive into the first four of the listed grantee projects, presenting an organizational overview, the context of their project, the activities they explored, and which of those aligned with the *Enterprise Criteria*. The organizations are as follows:

- **The American Indian Community Housing Organization** in Duluth, Minnesota: revitalizing a rooftop garden as a native community collaborative space with green infrastructure and traditional foods.
- **The Center for Neighborhood Technology** in Chicago, Illinois: creating a social and environmental justice initiative with local partners, developing four site-specific art and green infrastructure installations within a half-mile of transit stops in areas of high economic hardship.
- **Chinatown Community Development Center** in San Francisco (SF), California: enhancing social cohesion and climate resilience by building government partnerships in redesigning Portsmouth Square Park and the implementation of an ecofair.
- **Coalfield Development Corporation** in Wayne, West Virginia: providing out-of-work coal miners with retraining in reforestation, solar installation, furniture making, and sustainable agriculture on former mountaintop removal sites.

American Indian Community Housing Organization (Duluth, Minnesota)

AICHO was founded in 1993 to create culturally specific programming for Native women and children escaping violence. Since the opening of their Dabinoo'igan Shelter in 1994, AICHO has expanded to include its GiiweMobile Team (providing housing subsidies and case management to 33 families and 16 units of permanent housing for individuals and families coming from homeless situations). In 2006, AICHO purchased and renovated a historic YWCA, developing 29 units of publicly supportive housing and an Urban Indian Center. Opened in 2012, Gimaajii-Mino-Bimaadizimin also houses tribal partners and hosts cultural events, art shows, and performances. According to the organization, the center has become the central hub for the Native community in the region, where Native people go for safety, services, and community. It has also become a place where Native and non-Native communities connect and communicate.

Duluth is located in northern Minnesota on the shores of Lake Superior, populated by 87,000 people (93% white). American Indians make up 2.4% of the total population, yet they are an overrepresented percentage of the homeless population and face a wide range of barriers related to housing, employment and education. Michelle LeBeau, Director of AICHO described how many of the residents they serve: "Many Native American people move to Duluth in search of opportunities they can't find on their own reservations or in Minnesota's rural towns. As a whole, however, Duluth has not been welcoming to its indigenous population and tensions arise quite often over the city's negligence when it comes to including the Native voice in urban planning and public arts initiatives" (AICHO staff member). AICHO defines their community as the Indigenous community, the adults and children who live in Gimaajii-Mino-Bimaadizimin ("We are, All of us Together, Beginning a Good Life"), as well as the Indigenous artists they work with. Their project is rooted in Indigenous values but encourages participation from the greater Duluth community, especially local children.

Project description

In this project AICHO is responding to the challenges of climate by generating clean energy, improving access to traditional foods and medicines, and reducing water run-off by expanding use of the roof at the Gimaajii-Mino-Bimaadizim Urban Indian Center and redeveloping it as a rooftop garden and community learning and gathering space. They seek to engage Indigenous leaders, elders, artists, and community members to teach about traditional medicines, food, and cultural practices related to protecting the environment through cultural hands on experiences that address climate and cultural resilience practices.

Challenges addressed

- Lack of inclusiveness and knowledge of American Indian (AI) culture in local community;
- High levels of food insecurity for local AI population;
- Persistent power outages with increased severe winter weather events.

Proposed goals

Rooftop garden redesign and capacity-building for sustainability efforts:

- Installation of a 12 kW photovoltaic array;
- Installation of composting system;
- Installation of water capture system;
- Public art;
- Creation of a C&CR speaker series;
- Development of a C&CR Community Committee.

Process/design strategy

Prior to installing any climate infrastructure, AICHO had a community meeting to determine the type of programing and art that the community would want to see. AICHO partnered with Honor the Earth, a nonprofit environmental conservation organization, and Mayan artist Volton Ik, with the assistance of Derek Brown of the Dine or Navajo tribe, to design and paint a mural on the rooftop. The artists designed and painted the piece, but some residents of the building assisted in painting. Once it was complete, AICHO then had a community event with a cultural fair to unveil the mural. The event had drum circles, presentations about local events, and prayers for the community space by a tribal elder. The cultural fair included booths discussing climate and American Indian cultural topics. Even I had a table facilitating a participatory activity where community members mapped different things they wanted to see in the rooftop community garden space. These maps were used to inform the design for the rooftop space.

As people lined the street to watch the unveiling of the mural you could feel the excitement in the air. A dark sheet dropped to reveal a 30-ft portrait of an American Indian woman. On her body she wore a jingle dress, a dress traditionally worn by Ojibwe women during powwows that, when danced in, creates an airy, jingling sound, and covering her face a red bandana to represent the women who "participated in the Zapatista uprising in the Mexican state of Chiapas in 1994" as well as those "Water Protectors at Standing Rock." From the crowd, I heard sighs, cheers, and even sniffles trying to hold back tears. I knew immediately that this was more than a mural. In a building that houses Native women from some of the most difficult circumstances, in a city that has no public art for and by indigenous people, this was a symbol of resilience. "The mural is powerful in a spiritually and deep-down emotional way. Its interconnections with the importance of protecting our water, indigenous women, and cultural ways will now be front and center in downtown Duluth," said Ivy Vainio of the Grand Portage Band of Ojibwe. Vainio is program coordinator at AICHO.

Chicago Connections (Chi-Go) (Chicago, Illinois)

Chicago Connections is a coalition of six nonprofit organizations across the city. Four are local leaders in their respective communities, and two provide community-based art, storm water management, and technical and analytical support to all communities. All share a history of collaboration and a dedication to equitable and sustainable community development. The project lead is Center for Neighborhood Technology (CNT), a four-decade-old nonprofit dedicated to urban environments that are resilient, sustainable, and livable for people from all walks of life. CNT provides analytics and innovation, with a long history in testing, promoting, and facilitating implementation of economically efficient environmentally sound solutions.

The project co-lead is Arts + Public Life, an initiative of U Chicago Arts. Arts + Public Life builds creative connections on Chicago's South Side through artist residencies, arts education, and artist-led projects and events.

The 51st Street Green Line lead nonprofit is Urban Juncture, whose tagline is indicative of their mission—"Where Commerce Meets Community." Urban Juncture serves the Bronzeville community, where it develops commercial real estate and related enterprises addressing the needs of underserved communities.

The Blue Line Logan Square lead nonprofit is LUCHA, the Latin United Community Housing Association. Residents of Humboldt Park, West Town and Logan Square founded LUCHA in 1982, to combat displacement and preserve affordable housing in the community. LUCHA's current work includes the building of affordable housing developments as well as helping families rent decent and affordable housing.

The Pink Line California lead nonprofit is the Open Center for the Arts (The Center), which provide a space where all artists can come together to educate, showcase, refine, and develop their talents.

The Blue Line Homan Square lead nonprofit is the School of the Art Institute (SAI). Through its 7000 Oaks for Chicago project, SAI brought artists to its mobile foundry in its North Lawndale Homan Square campus, holding community events, art programming, and planting of trees.

The four communities face multiple challenges, including high levels of economic hardship, urban flooding, and high combined housing and transportation costs for residents. Equity issues are also a concern, as the areas face questions of safety and disinvestment. The four community areas are designated by Chicago Department of Public Health 2014 statistics as "high" levels of economic hardship. Each location has distinct cultural histories and populations, and because of the segregation across the city, collaboration and connection is needed but not the norm. These collaborators recognized the importance of pooling their resources to have a more collective and impactful effort for the residents of Chicago.

Project description

Chi-Go is a social and environmental justice initiative that aims to strengthen social networks and climate resilience through public arts and storm water projects. This project is unique in that they are working to connect four neighborhoods across Chicago, each of which has a very different experience in the city. Chi-Go facilitates cross-cultural learning between each of the four areas. For example, the lead nonprofit groups are coming together to share their approach, processes, barriers, and results to support one another. The two convening nonprofits—CNT and Arts and Public Life—are documenting the results and leading the production of a final report.

Challenges addressed

- Chicago is a city divided. In 2015 it was the most segregated city in the country;
- Dominated by hardscape and regular rain causes severe flooding;
- Severe threat of gentrification.

Project goals

- Four artists/teams selected;
- Youth engaged in installation;
- Four installations completed;
- Four or more kickoff events/celebrations;
- Best practices doc completed, capturing process of art installation and community responses.

Process/design strategy

Chi-GO is engaging in specific place-based initiatives in each of their four areas, which build upon existing community assets and resident expertise around train stops. All of their activities are linked through a common arts participation and a stormwater management strategy, which engages community residents and stakeholders. Each area has a lead partner nonprofit that works to: (1) involve local artists; (2) connect to local anchor institutions such as schools and houses of worship; and (3) direct local implementation efforts with community residents and stakeholders. A community-led committee selected the local artists. The artists then engaged in multiple community participatory sessions to identify and develop public art installations, programming, and community gardens.

In Chicago's Cook County gun violence resulted in more than 744 deaths in 2016—more homicides and shooting victims than New York City and Los Angeles combined. Community members of the Holman Square neighborhood planted its first 10 of 7000 trees in a community event where local artisans melted down reclaimed guns from a gun amnesty program and repurposed them into shovels. The opportunity to get involved in the community while opening up communication around the very pressing resilience challenge of gun violence was groundbreaking (literally and figuratively). This event is part of a larger project to continue planting trees across the neighborhood and could raise ultimately increase canopy coverage to more than 38% over current canopy amounts decrease surface temperatures, reduce heating and cooling costs for residents by \$38–\$77 per household annually and, once mature, reduce the frequency of some crimes in the 24th ward by up to 7.7%. This would result in crime avoidance savings of \$1.3 million annually. For Chi-Go, this work has been integral to strengthening social networks within and across these diverse communities, especially within African-American and Latino communities.

Chinatown Community Development Center (San Francisco, California)

Founded in 1977, the mission of the Chinatown Community Development Center is "to build community and enhance the quality of life for San Francisco residents" (CCDC, 2018). They are a place-based community development organization serving primarily the Chinatown neighborhood (CCDC, 2018). They not only maintain affordable housing, but also plan and rebuild neighborhood parks and alleyways and strengthen and protect Chinatown small businesses and restaurants. San Francisco's Chinatown has a population of 18,000, largely foreign-born Chinese immigrants who are slightly older than the city average. Residents' poverty rate is almost three times the San Francisco average; their unemployment rate is over two times the average. Sixty percent of Chinatown families and seniors live in overcrowded single room occupancy (SRO) residences. According to the SRO Families Collaborative 2015 Census, there are 530 SRO buildings in San Francisco's Chinatown. SROs, built to house bachelor Chinese laborers in the past, do not adequately accommodate families. As the cost of housing in the city soars, more families are forced to live in these cramped units. SROs house an estimated 457 families with children and, among these families, 62% immigrated from China or Hong Kong. Only 14% of SRO heads of household speak fluent English.

In SROs, parents raise children in rooms that typically measure 8 by 10 ft, about the size of a large walk-in closet. These units are crammed with bunk beds, desks, dishware, and clothing, leaving little space for residents to do more than sleep. This overcrowding frequently results in physical and mental health challenges. Despite the challenging conditions, the alternative is homelessness. The dual solution is stabilized housing so families can become more self-sufficient, as well as adequate outdoor space that serves as a respite from these crowded SROs. Chinatown is an immigrant gateway and serves as the current and historic heart of the social and cultural life of the Chinese-American and Asian Pacific Islander (API) communities in the region. It is a destination for millions of visitors each year. The Sustainable Chinatown project has the potential to reach everyone who lives, works, plays or visits the neighborhood and its assets.

Project description

This project improves the Chinatown neighborhood's environmental performance and sustains the community's unique culture and history with a Chinatown Eco Fair and a Portsmouth Square park community engaged redesign.

Challenges addressed

- Lowest parks per person in city (11 vs 300 ft^2/person);
- Dominated by hardscape: very little tree cover and permeable surfaces;
- No disaster preparedness plan or place to evacuate in case of emergency;
- Threatened with rapid increase in urbanization and increased real estate prices.

Project goals

- Preservation of the Chinatown neighborhood;
- Develop a city plan that reflects the present and future needs of the community and positions Portsmouth;
- Square to receive funding for the resulting capital improvements;
- Develop a final conceptual design for the redeveloped Portsmouth Square Park.

Process/design strategy

Chinatown CDC recognized a gap in the knowledge and understanding of climate issues in the community. To share valuable information about climate threats in the neighborhood and interventions to overcome them, they organized an Eco Fair in Portsmouth Square park, led by high-school-aged youth in Chinatown, in August 2017. The fair featured conservation information and demonstrations, recycled art activities, and plant giveaways for those that visited all of the respective booths, each displaying important climate focused information. All information was translated into Chinese by the youth. The Eco Fair was used as a tool to creatively engage Chinatown residents around the vast and pressing climate issues impacting the neighborhood. It included a culturally relevant participatory activity that used a wishing tree for Chinatown residents to share their wishes for the future of Chinatown and Portsmouth Square park.

For the park redesign, they employed a community planning process for Portsmouth Square that included four community meetings, one-on-one interviews with stakeholders and intercept surveys with users of the park. Ultimately, their process identified sustainability strategies that include green infrastructure and a resilience center that can be useful for everyday life and in times of disaster. The Portsmouth Square redesign incorporates water and energy saving technologies and integrated green infrastructure strategies to improve neighborhood resilience, and improve community cohesion. CCDC is also implementing a public education campaign that teaches about water and energy savings, and the impacts of heat and other climate vulnerabilities on human health. It focuses on ways to prevent negative health impacts of a changing climate in a neighborhood that has a high urban heat island effect. This effort is bringing community members together to learn about topics they are less familiar with in order to advocate for a community informed redesign. Residents are more knowledgeable, engaged and invested in the development processes they may typically be excluded from.

Coalfield Development Corporation (Wayne, West Virginia)

Coalfield Development provides out-of-work coal miners with retraining in reforestation, solar installation, furniture making, and sustainable agriculture on former mountaintop removal sites. Coalfield's vision is to develop the potential of Appalachian places and people as they experience challenging moments of economic transition. Since 2009, after significant community engagement, Coalfield pioneered a relationship-based, holistic approach to on-the-job training. They hired unemployed and underemployed people to construct green affordable housing. Trainees worked the 33-6-3 model each week: 33 h of paid labor, 6 h of higher education class time, and 3 h of life-skills mentorship. Today, they have grown into a collection of social enterprises working throughout Appalachia to create a more sustainable economy in the wake of the coal industry's rapid decline. They have created more than 40 on-the-job training positions, more than 200 professional certification opportunities, redeveloped more than 150,000 ft^2 of dilapidated property, and successfully launched five new businesses in real estate development, construction, wood working, agriculture and artisan trades—all of which are industries based on local assets in the Appalachian region.

Mingo County is a rural county in southern West Virginia with a total population of 25,900. This project serves those making less than 50% Annual Median Income (AMI). According to the Energy Information Administration, total coal output from West Virginia underground and surface mine operations fell to 113 million short tons in 2013, marking the lowest amount produced in the state since the early 1980s. Mingo County has been distressed by the downturn in the coal industry. As consumers turn to greener, more sustainable forms of electricity, the coal industry has essentially collapsed. While socioeconomic indicators have always lagged national averages, the situation has declined rapidly: those living in poverty are 28.1% (compared to 13.5% nationally), unemployment is 13.4% (4.8% nationally), labor participation is only 49.1% (62.7% nationally), and per- capita-income of $32,902 ($55,966 nationally) (Arc, 2018). The economic hardship contributes to social hardship. Only 16.4% of Mingo residents have a degree of higher education, compared to 37.2% nationally. Rates of cancer in West Virginia are alarmingly high: 194.4 new cases per 100,000 people. Only Kentucky and

Mississippi have higher rates (Asco, 2018). Deforestation caused by mountain-top-removal coal mining has ruined over 1200 miles of streams, flattening 500 mountains, and decimating 1.2 million acres of forest in Appalachia (Arc, 2018).

Project description

This project retrains out-of-work coal miners in reforestation, solar installation, furniture making, and sustainable agriculture on former mountain top removal sites and does so by combining job-training, creative placemaking, and culturally grounded mentorship with reforestation of former mine sites.

Challenges addressed

- Rich coal heritage in sharp decline;
- Lack of economic diversification and underemployment;
- Mountaintop removal has deforested hundreds of acres of land.

Project goals

- Retrain unemployed coalminers in green and culturally support jobs;
- Revitalize the economy;
- Increase local pride;
- Shift identity association from sole coal focus.

Process/design strategy

In addition to reclaiming the scarred landscape, this project enables laid-off coalminers and other low-income Appalachians to rediscover their land and culture and to appreciate the local craftsmanship, artistry, music, and culture. Local artists and artisans mentor low-income trainees in trades unique to Appalachia: quilting, woodworking, mountain music, foraging/canning, beekeeping, pottery, and glassblowing. These same trainees are hired in partnership with TNC to do reforestation work, solar installation, sustainable agriculture, and green housing on former mountain-top-removal sites.

They are currently testing sustainable agriculture and reforestation on former mountain-top-removal sites- achieving the creative placemaking principles directly in places that have created the most devastation to the land and its people. Each trainee devotes 3 h a week to life-skills mentorship. During the 3 h, local artists and artisans connect creativity to life skills such as problem solving, teamwork, communication, personal health, and even financial literacy. Because of the 33-6-3 model, formerly unemployed people gain employment and a renewed self-confidence. Cultural expression is enhanced over the course of the 2-year contract granted to each trainee. By completing transformative community projects, a greater sense of community ownership manifests. Coalfield is creating an opportunity for former coal miners to reinvent themselves to be resilient, while maintaining their identity.

From building the case study through focus groups, interviews, observations, existing data and research, and my own participant observations, I uncovered the following.

4.2 Understanding climate and cultural resilience

There are alignments and misalignments in understanding climate and cultural resilience by community members and clear prioritization of other more critical issues, many of which communities attempted to incorporate in the grant with perhaps less success. It is evident that while there are similarities across communities, the types of resilience challenges most important to communities are diverse and context-specific. Fig. 2.3 shows clearly that creative placemaking processes across the board were able to facilitate *Enterprise*'s community-engaged activities that focused on cultural expression and local partnerships with culture bearers. The next most common activities were creating partnerships with artists to create a product reflecting the community's identity and creating a plan for addressing climate resilience needs coupled with cultural resilience efforts for added impact.

From the focus groups data, art, and creative and cultural practices support three main outcomes across grantee groups. Those outcomes are lifting resident voice and identity, empowering residents to feel pride and support in place, and fostering social connection and communication. All three of these outcomes are cultural-resilience-focused. However, fostering social connections and communication, as mentioned in the literature, is a shared outcome that supports both climate and cultural resilience. There is a gap in knowledge for community development organizations in fully understanding this connection and being able to use it to plan and support for more impactful climate programming. Discussions of increased safety, political involvement and beautification were other underlying themes but were not highlighted across communities.

	AICHO	WonderRoot & Southface	Chi-Go	Chinatown	Coalfield
(1A) Learn and share with residents about climate impact		I		I	
(1B) Proactively address climate challenges	I		I	I	
(1C) Plan to address climate resilience needs to be paired with cultural resilience efforts for added impact	I		I	I	I
(2A) Partner w/ artists/ designers and community to create product reflecting community identity	I	I	I	I	I
(2B) Conduct community engaged activities that focus on cultural expression of people involved	I	I	I	I	I
(2C) Use culturally competent practices to deliver services and gain stakeholder input	I			I	

FIG. 2.3 Enterprise C&CR activities criteria. *(Source: Venable-Thomas.)*

4.3 Understanding creative placemaking

Creative placemaking is a concept not well understood on the ground. However, communities recognize the importance of creative activities in supporting community resilience. Additionally, the application of creative placemaking projects varied across communities with the achievement of many of the same outcomes, highlighting the importance of a few primary required creative placemaking activities. Much like understanding climate and cultural resilience, the creative placemaking interventions were most successful when focused on processes that best addressed place-based, community-expressed needs. Although all the creative placemaking practices were not exactly the same, there were several components that were shared across grantee groups. All of the grantee organizations used some sort of community participatory activity to engage and incorporate community voices into the project. Every project engaged other sectors and supported interdisciplinary partnerships to implement their projects. All of the projects highlighted some social issue in their community, such as violence or discrimination. Lastly, every intervention was place-based and publicly visible. As pictured in Fig. 2.4, these

	AICHO	WonderRoot & Southface	Chi-Go	Chinatown	Coalfield
Community Commitee for decision making	I	I	I		
Artist led installation	I				
Community led installation			I		I
Community participatory design activity/design charettes	I	I	I	I	I
Cross sector partnerships/ local insitutions/community groups	I	I	I	I	I
Incorporates social issues	I	I	I	I	I
Place-based interventions	I	I	I	I	I
Publicly visible interventions	I	I	I	I	I
Equal community partnership		I	I	I	
Local artist/designer		I	I		I

FIG. 2.4 Creative placemaking approach criteria. *(Source: Venable-Thomas.)*

different criteria demonstrate that creative placemaking interventions can vary across different communities but there are some baseline criteria for creative placemaking interventions that can support certain community resilience outcomes. Creative placemaking supported producing climate and cultural resilience activities as well as producing community social outcomes. However, it did not effectively address many of the other more pressing issues in communities, such as trauma and displacement, and in some places was thought to spur them.

5 Discussion

This program revealed that cultural resilience is not only integral to the ways in which communities understand and experience community resilience but an approach that supports increased social connection of residents. Although demonstrated in diverse ways across grantee communities, a shared mechanism for resilience building was the use of culture and creativity to connect within groups, and to innovatively overcome challenges in place. Creativity and cultural practices were discussed as tools or a protective measure against trauma and other shocks and stressors that resident's experienced. For example, the common theme across AICHO's spaces including the first floor and lobby of their supportive housing unit that serves over 33 families was of American Indigenous cultural representation. For residents and others that frequent the shop and or community center, it was evident that native art was centered there, and this was also reflected in the responses heard from community. "Culture is collective and it's based in your understanding of the now. Sometimes we think of culture as historical—that's DNA stuff, like our ancestral knowledge. It's like, how do we connect and that learning is what's going to be there. We do that through arts" (AICHO resident). Although creative placemaking was not an expression that community members used, they understood the value of art and creative expression in connecting residents to each other and residents to staff. A Chi-Go resident described the connection between culture and creative practices and language. "Culture and creativity—a lot of people use them as common language. The language of helping one another" (Chi-go resident). The residents at AICHO also felt that the art brought people together because it reminded them of their shared culture that is sometimes hard to find when there are so many other challenges. "Drum leagues, choirs, dance troops, arts are helping keep the history and cultural needs [of the community] alive [but] other challenges have started to overwhelm/overshadow the work they are doing (AICHO resident)." Creative placemaking is a great start to highlighting the existing work of the community, sharing stories, and creating spaces to collaborate but it has limitations.

Because underserved communities in the United States often experience disinvestment and inequity because of the cultural groups that they identify with, many of their resilience challenges are inadvertently connected to their cultural identities. Climate and culture alone do not fully conceptualize the community resilience model. There are even more components than anticipated from this program that are important to a community's ability to be resilient, such as economic and social issues on top of the experiences of preexisting trauma. This program demonstrated that resilience is not only about rebounding from environmental risks but also about persevering through traumatic experiences and other perpetually stressful events. As an intermediary organization whose role is to collaborate with government and nongovernment organizations while focusing on development, it is important for *Enterprise* to understand how utilizing a resilience paradigm might augment traditional efforts to prepare communities to withstand anticipated disasters and emergent and consistent threats (Yosso, 2005). This lies in understanding the significance of human health, wellness, and culture to overall community resilience. Incorporating creativity and cultural identity into community participatory planning processes shifts the research lens away from a deficit view of underrepresented communities, and instead focuses on and learns from the array of cultural knowledge possessed by socially marginalized groups that often go unacknowledged. In addressing resilience issues in underrepresented communities, culture and creativity coupled with community participatory processes must be integrated in public health, environmental, urban planning, and development research and programs, to build institutional knowledge and practice.

Based on my findings, my recommendations are that organizations must be adaptive to redefining community resilience and changing the language around creative placemaking while creating shared metrics for evaluating impact. As seen in Fig. 2.5, community resilience needs to incorporate more components than climate and cultural resilience such as economic resilience, social connection and capital, and healing centered responses that are closely tied with social determinants of health. This distinction provides a more accurate conceptual framework for understanding resilience that further supports the connection between creative placemaking, culture and community resilience. Additionally, it further connects this resilience work to other disciplines like public health doing the same work. This broadens the field of literature, practice, and expertise that can be accessed to advance resilience work. The language around creative placemaking needs to better incorporate place-based context. Organizations need shared metrics that cross disciplines to begin evaluating the impact of and interaction between creative placemaking and community resilience efforts. I again presented a tool that creates an indicator model leveraging the outcomes uncovered in this program, connecting them to evidence-based indicators, and

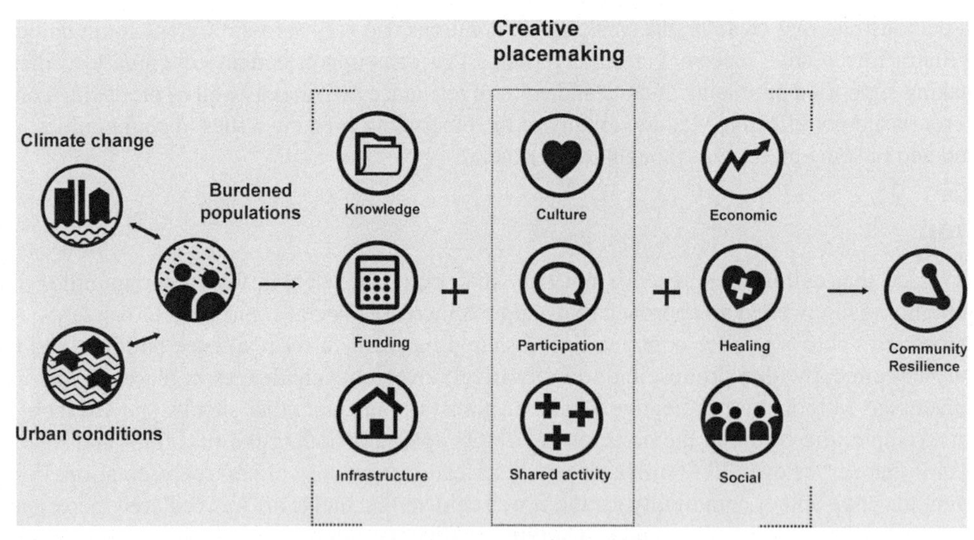

FIG. 2.5 Revised conceptual framework for community resilience. *(Source: Venable-Thomas.)*

putting them into a survey tool that can be used at the project level for evaluation. More resources need to focus on better understanding how well creative placemaking is addressing cultural resilience and what the true impact of cultural resilience is on overall community resilience.

6 Conclusions

My goal was to investigate how climate resilience, cultural resilience and creative placemaking are understood among different stakeholders engaged in community development, the role of creative placemaking in advancing climate and cultural resilience, and the role that an intermediary might be best suited to influence these strategies. From this research, it is clear that cultural resilience is a component of community resilience that if not included leaves significant value on the table for community development organizations and their funders. By exploring five different exemplars for using creative placemaking as a tool for building community resilience and analyzing their alignment with expressed community needs I was able to uncover how the communities of the C&CR Program were conceptualizing the concepts of creative placemaking and community resilience and if, in fact, these conceptualizations aligned across stakeholders. In order to evaluate the impact of these concepts it was important to first understand the ways in which these concepts were realized in practice. Through focus groups, interviews, observations, existing data and research, and my own participant observations, I created a case study of these five exemplar organizations participating in the C&CR Program. There must be an understanding that creative placemaking cannot solve all acute or chronic stressors, particularly because the issues exacerbated by the impacts of climate change are those created by community disinvestment over time. Creative placemaking can be a mechanism for bringing diverse voices to the table. Recognizing the importance of cultural resilience, as a viable approach for community development and community building is extremely important because research, health, and community development decisions are still being made without the inclusion or voice of those most impacted by the decisions. We are seeing that this is a mechanism for inclusiveness. When we think of resilience as a privilege unequally supported across different communities, it changes the responsibility of stakeholders in investing in equitable interventions.

Acknowledgments

This work would not have been possible without the Doctorate in Public Health Program at Harvard T.H. Chan, which has allowed me to push the boundaries of what a public health professional can do. I must also thank the members of my *Enterprise* team for making this project happen and allowing me to be a part of this journey. I would like to thank my parents, whose love and guidance are with me in whatever I pursue. They are the ultimate role models.

References

Acosta, J.D., Chandra, A., Madrigano, J., 2018. An agenda to advance integrative resilience research and practice: key themes from a resilience roundtable. RAND Corporation. [Report] Retrieved from: https://www.rand.org/content/dam/rand/pubs/research_reports/RR1600/RR1683/RAND_RR1683.pdf.

Aldrich, D.P., 2012. Building Resilience: Social Capital in Post-Disaster Recovery. University of Chicago Press, Chicago, IL.

Arc, 2018. The Appalachian Region: A Data Overview From the 2011–2015 American Community Survey. Appalachian Regional Commission. Retrieved from: https://www.arc.gov/research/researchreportdetails.asp?REPORT_ID=132.

Ascopost.com. (Asco), 2018. Stemming the Growing Cancer Crisis in Rural Appalachia. The ASCO Post. Retrieved from: http://www.ascopost.com/issues/september-25-2017/stemming-the-growing-cancer-crisis-in-rural-appalachia/.

Berkman, L., Kawachi, I., 2001. Social ties and mental health. J. Urban Health 78, 458–467.

Berkman, L., Kawachi, I., Glymour, M., 2015. Social Epidemiology. Oxford University Press, New York, NY.

Chinatowncdc (CCDC), 2018. Retrieved from: http://www.chinatowncdc.org/about-us.

Clauss-Ehlers, C.S., 2010. Cultural resilience. In: Clauss-Ehlers, C.S. (Ed.), Encyclopedia of Cross-Cultural School Psychology. Springer, Boston, MA, pp. 324–326.

Community-Wealth (CW), 2018. Community Development Corporations (CDCs). Retrieved from: https://community-wealth.org/strategies/panel/cdcs/index.html.

Crabtree, B., Miller, W., 1992. Doing Qualitative Research. Multiple Strategies. Sage, Thousand Oaks, CA.

Howard, D.E., 2003. How alienation and declines in social capital are affecting adolescent well-being. Working Paper, Department of Public and Community Health, University of Maryland, College Park, MD.

Kawachi, I., Kennedy, B., Lochner, K., Prothrow-Stith, D., 1997. Social capital, income inequality, and mortality. Am. J. Public Health 87, 1491–1498.

Lalonde, C., 2005. Identity formation and cultural resilience in aboriginal communities. In: Flynn, R.J., Dudding, P., Barber, J. (Eds.), Promoting Resilience in Child Welfare. University of Ottawa Press, Ottawa, pp. 52–72.

Putnam, R., 2000. Bowling Alone: The Collapse and Revival of American Community. Simon and Schuster, New York, NY.

Searle, J., 1995. The Construction of Social Reality. Free Press, New York, NY.

Stake, R.E., 1995. The Art of Case Study Research. Sage, Thousand Oaks, CA.

Yin, R.K., 2003. Case Study Research: Design and Methods, third ed. Sage, Thousand Oaks, CA.

Yosso, T., 2005. Whose culture has capital? A critical race theory discussion of community cultural wealth. Race Ethn. Educ. 8 (1), 69–91.

Young, N., 2017. Enterprise Community Partners Climate and Cultural Resilience Request for Proposal.

Chapter 3

Mainstreaming cultural heritage in disaster risk governance

Rohit Jigyasu
International Centre for the Study of the Preservation and Restoration of Cultural Property (ICCROM), Rome, Italy

1 Impact of disasters on cultural heritage

Our understanding of cultural heritage has undergone a marked shift during the last few decades in terms of what it is, its significance and ways of protection and management. Cultural heritage today encompasses various typologies including not just monuments and archaeological sites but also historic cities, cultural landscapes as well as industrial achievements of the past and even places that are associated with painful memories and war. Moreover, collections of movable items within sites, museums, historic buildings, libraries and archives are linked not only to significant artistic and literary developments but also to day to day life of the people. At the same time, intangibles such as knowledge, beliefs and value systems are fundamental aspects of heritage that determine people's daily choices and behavior. As such cultural heritage continues to perform its important role as a source of identity for communities, who are true bearers of heritage.

However, each year disasters caused by natural and human induced hazards cause enormous damage to cultural heritage, depriving communities of their irreplaceable cultural assets. Damages to cultural landscapes and local flora and fauna cause loss of ecosystem services thereby putting the sustainability of local communities at risk. Moreover, disasters also adversely affect intangible cultural heritage represented by traditional knowledge, practices, skills and crafts that ensure cultural continuity, as well as the means for its protection and maintenance.

There have been many recent global examples that demonstrate the impact of disasters on cultural heritage properties. In 2019, fires ravaged the World Heritage Sites of Notre Dame Cathedral in Paris, France, and Shuri Castle in Okinawa, Japan. The previous year, a massive fire devastated the main building of the national museum of Rio de Janeiro, Brazil, thereby extensively damaging most of its collections. Earthquakes have also caused damage to cultural heritage as in the case of Nepal and Mynamar earthquakes in 2015 and 2016 respectively.

Climate change is increasing the number of disasters and their devastating impacts on cultural heritage. From 1988 to 2007, 76% of all disaster events were hydro-meteorological in nature (UNISDR, 2008). Due to high intensity rainfall, increased instances of urban flooding have been reported in recent decades in nearly every part of the world, inundating many historic centers such as Venice, Italy in 2019 or Passau, Germany due to floods in large parts of Central and Eastern Europe in 2013. The same year (2013), flash floods due to unprecedented heavy rains in India's Uttarakhand State destroyed many heritage structures in the region, and heavy rains in Thailand in 2011 caused the World Heritage Site of Ayutthaya to remain submerged in water thereby causing insurmountable loss to the foundations of historic built structures. More recently in 2018, massive floods in Kerala state in South India caused extensive damage to movable and intangible heritage characterized by manuscripts, traditional boats and the metal mirror industry. The likelihood of increased weather extremes such as heavy downpours, heat waves, and strong hurricanes and cyclones therefore gives great concern that the number or scale of weather-related disasters will also increase, putting cultural heritage at grave risk (Seneviratne et al., 2012). Moreover, climate change is also contributing toward removing and altering animals' habitat, changing how they live, where they live and who eats whom, and with increased urbanization, humans are more susceptible to pathogens carried by wild animals as has been the case of current Covid-19 Pandemic (Benton, 2020), which has also led to closure of many heritage sites and museums and halted festivals, cultural practices and traditional livelihoods.

2 Governance for disaster risk management of cultural heritage

To address these challenges, prevention, mitigation and preparedness measures, emergency response procedures, and recovery and rehabilitation should be undertaken as part of disaster risk management for cultural heritage before, during

Strengthening Disaster Risk Governance to Manage Disaster Risk. https://doi.org/10.1016/B978-0-12-818750-0.00003-9

and after any disaster. However, investing in disaster risk reduction through mitigation and preparedness makes much more economic sense than investing heavily on response and recovery, as previous experiences in post disaster recovery of cultural heritage in Nepal, Myanmar and Italy have aptly demonstrated. Going by the widely accepted principle of "Building Back Better," the recovery and rehabilitation process should incorporate mitigation of risks for future disasters with minimal impact on heritage values.

On one hand, this would necessitate each heritage site and museum to have its own disaster risk management plan that is tailored to its specific characteristics. On the other hand, cultural heritage needs to be well integrated into overall disaster risk management policies and plans at national, regional and local levels.

Governance and management are closely related concepts but to some extent it is useful to distinguish them. While governance is fundamentally about who takes decisions and how those decisions are made, management is about what is done to implement those decisions and the means and actions needed to achieve them (Borrini-Feyerabend et al., 2013). Therefore, governance is about who decides, what the objectives are, what to do to pursue them and with what means. How those decisions are taken? Who holds power, authority and responsibility? Who is (or should be) held accountable? It is also important to mention here that the line between management and governance will vary from one situation to another (Booker and Franks, 2019).

This distinction between governance and management is also helpful as it highlights the need to consider how disaster risk management of cultural heritage is influenced by economic, social and political considerations and not just technical ones. Interestingly both disaster risk management and heritage conservation have seen a paradigm shift from predominant techno-centric focus to more attention to policy and power related aspects recognizing the importance of robust governance systems than ever before.

Often governance is inaccurately used as a synonym for government (Graham et al., 2003). Where the later defines a set of institutions responsible for administrating a country or state, the term governance can be used in a broader sense and applied to a wider range of circumstances. We can talk about the governance of a financial market, the governance of a company as well as the governance of a heritage site or other type of territory. Governance can be undertaken by government (at various levels), various rights-holders and stakeholders together (shared governance), private individuals and organizations, and indigenous people and/or local communities (Borrini-Feyerabend and Hill, 2015).

Each time choices and decisions are made about disaster risk management for a cultural heritage, some forms of "governance" systems exist, which enable people to exercise their rights, influence, authority and responsibilities over that place. Thus, understanding *who* takes those decisions and *how* they make them is important to assess how well a specific cultural heritage is being protected and disaster risks mitigated. The units of governance for cultural heritage as well as disaster risk management are usually determined on specific criteria, although these may fall within larger administrative units. Also, often multiple units of governance may overlap with each other in case of larger heritage sites such as cities and cultural landscapes.

3 Who makes decisions?

Disaster risks governance for cultural heritage involves many actors, who are not solely tied to the institutions of government. Actors may also include elected and traditional authorities, indigenous peoples and local communities, private owners, businesses, non-profit trusts, NGOs and international agencies, professional organizations, religious and educational organizations, etc. Such actors hold rights, influence, authority and responsibilities over cultural heritage and disaster risk management through laws, plans, norms, traditions and other similar instruments, which determine their roles and powers at their disposal.

Three broad categories of actors can be distinguished: management agencies; rightsholders and stakeholders. Management agencies refer to institutions and other types of entities which are recognized, responsible and accountable for protecting and managing cultural heritage and disaster risks at various levels. Rightsholders are socially endowed with legal or customary rights over cultural heritage but are not directly responsible for its management. Stakeholders possess direct or indirect interests, concerns and influence over cultural heritage and disaster risk management but do not necessarily enjoy a legally or socially recognized entitlement (Borrini-Feyerabend et al., 2013).

4 Need for coordination between cultural heritage and DRM sectors

In order to undertake effective measures for disaster risk reduction of cultural heritage, there needs to be greater cooperation between agencies and professionals from heritage and disaster management fields. For emergency response and post disaster recovery, heritage professionals and agencies should work closely with civic defense organizations as well as local, national and international organizations engaged in recovery and rehabilitation. Until a few years ago, disaster risk managers would shrug off cultural heritage as an elitist sector and therefore it was largely believed that disaster risk management sector could not afford to indulge in it, when life, assets and livelihoods are the primary concerns. However, over the years, it has been shown that cultural heritage in fact contributes in many ways to reducing disaster risks as well as responding and recovering, but much more needs to change on ground.

Sendai Framework on Disaster Risk Reduction adopted in 2015 clearly recognizes culture as a key dimension of disaster risk reduction and the need to protect and draw on heritage as an asset for resilience through a number of important references. The challenge is to implement this policy, which requires considerable building of capacities at international, national and local levels and the setting up of the necessary institutional mechanisms, complemented by data collection and monitoring (Dean and Boccardi, 2015). The framework has clearly advocated reducing risks to cultural heritage in the national policies on disaster risk management.

On the other hand, disaster risk management for cultural heritage needs to be integrated into various development sectors such as sanitation, water supply, housing, environment, infrastructure and services. Unfortunately, on the ground there is very little coordination between cultural heritage, disaster risk management and development agencies or departments. This can be exemplified through the case of Jaisalmer Fort in Rajasthan, India, which is on UNESCO's World Heritage List. This fort is inhabited by people since centuries. Heavy rains in 2014 lead to the collapse of the fort wall. This was due to leakage of water supply pipes that were installed by the municipality in the absence of any coordination with Archaeological Survey of India that is responsible for protecting the built heritage. In case of urban heritage, often there are lack of legislations and/or policies for the risk-sensitive protection of cultural heritage assets in a city and heritage concerns are not included in urban vulnerability reduction programs.

To effectively reduce disaster risks to cultural heritage, agencies responsible for heritage conservation and management should be able to integrate disaster risk management within their site management procedures and practices. On the other hand, organizations responsible for disaster management should be able to include heritage concerns within mitigation, preparedness, response and recovery strategies. Moreover enhanced communication between site managers or those managing cultural institutions such as museums and archives with departments/institutions forecasting warning of flooding, rainfall, cyclones would help the former prepare before disaster strikes.

This would necessitate building capacity at various levels and among various organizations, but most importantly facilitate interaction between decision makers, professionals and managers from heritage, disaster management and development sectors. This would help them understand each other's vocabulary and tools and hold better dialogue and coordination which is critical for effective disaster risk governance.

Past experience also shows that in the aftermath of a disaster, cultural heritage often gets destroyed due to uninformed action of national and international rescue and relief agencies, who demolish these structures in the absence of coordination with heritage management agencies and absence of proper methodology for damage assessment that takes into consideration both safety as well as heritage values. Often standard principles for contemporary "engineered" buildings are applied on historic and traditional "non-engineered" buildings with the result that many of them are categorized as unsafe and therefore worthy of demolition. Therefore, it is important that in pre-disaster phase, heritage is placed in the chain of command by ensuring that heritage expertise is present on relief teams, giving sufficient authority to heritage experts and establishing written protocols defining commitment to respect heritage.

Last but not the least, coordination between disaster risk management and cultural heritage agencies is needed for sharing of information, data and maps related to risks and vulnerabilities before a disaster, and damage and losses in post disaster situation, both at macro and micro levels. This calls for common definitions, protocols and standard formats for collecting and storing disaster loss data on cultural heritage at the national level through involvement of statistical agencies and other relevant organizations. Through such collaboration, cultural heritage can be included as one of the layers in order to assess specific risks to different components based on their vulnerability and exposure.

5 Utilizing traditional governance mechanisms

Institutionalized formal disaster risk governance systems are not always effective for managing disaster risks of cultural heritage. Rather traditional governance systems represented by the social and cultural mechanisms of communities not only mitigate the effects of disasters but also cope and aid in physical, social and psychological recovery. These mechanisms are embodied in mutual support systems that are determined by traditional relationships.

Traditional social and religious networks and management systems have been very effective in community led initiatives for disaster risk management for cultural heritage as exemplified in the case of World Heritage of Shirakawa-Gu Villages in Japan, where these have been successfully employed for monitoring and responding to the risk of fire.

In fact, traditional governance systems also have tremendous potential in securing collective action among communities for response and recovery following disasters. The rich expression of heritage is also a powerful means to help victims recover from the psychological impact of the disaster. In such situations, people search desperately for identity and self-esteem. Traditional social and religious networks that provide mutual support and access to collective assets, often represented by heritage, are an extremely effective coping mechanism for community members (UNISDR, 2013).

In many instances, traditional community-based heritage structures such as temples serve as anchors for these networks. Following the great East Japan earthquake and Tsunami in 2011, many victims especially school children could take refuge in historic temples that were located on higher grounds. Many such temples also served as shelter for the affected people for weeks and months and were supported by local religious and community leaders. Furthermore, they also enable social cohesion among communities thereby increasing their resilience and coping capacity against disasters. Following the 2011 Great East Japan disaster, the Shizugawa town in Minamisanriku-cho was almost completely swept away by the tsunami waves. However, the main hall of the local Kaminoyama Hachimangu Shrine survived the disaster due to its location on higher ground. The shrine, through its priest, acted as the local anchor for affected communities to get together to provide psycho-social support to each other and for sharing community views on recovery and reconstruction. The rituals, festivals and crafts were reinstated within a few weeks of the disaster, serving as means for mutual support among community members.

After 2015 Nepal earthquake, networks of traditional *guthi* (communal trusts) provided support to local communities in their transition from response to recovery phase. Traditional cultural practices also help in forging community solidarity, which is crucial for recovery afterwards. Take the case of traditional wood craftsmen in the earthquake affected area of Kathmandu, Nepal located in the World Heritage Monument Zones of Kathmandu valley: their traditional roles and responsibility were to cater to the building needs of the local community through their skills and they got returns for their service in kind (e.g., food grain). Social networks at joint family and neighborhood levels also helped communities to provide material and psychological support to cope with the trauma of a disaster. This is well exemplified in the case of Gujarat following the 2001 earthquake, where traditional community networks, religious and philanthropic institutional structures played a significant role in supporting post disaster recovery efforts. Even before external help could arrive, friends and neighbors came to help each other in immediate relief and rescue, even though each of them had lost their own family members and friends. For the initial few months, they were the best supporters and helpers. Their faith helped them gain internal strength to fight against the tragic circumstances and reassemble their disrupted lives. Such local coping mechanisms developed out of strong community and religious networks, also reflecting their psychological needs rather than mere physical ones, may result in many of the victims identifying the rebuilding of shrines as an urgent need in order to provide solace and support within families and communities (Jigyasu, 2002).

6 Effectively engaging communities

Therefore, close engagement of local communities living in and around heritage sites is crucial as they can contribute effectively toward disaster risk reduction of cultural heritage and also assist as volunteers during emergency situations. There is an increasing focus toward proactive and community-based approaches for disaster risk reduction. Many countries have adopted a proactive policy approach to disaster management that focuses on reducing risks—the risk of loss of life, economic loss and damage of property, especially to those sections of the population who are the most vulnerable due to poverty and general lack of resources. Some of these also aim to protect the environment. This approach involves a shift away from a perception that disasters are rare occurrences only to be managed by emergency rescue and support services.

However, contrary to larger perception, communities are not homogenous. Rather they can be socially differentiated and diverse. Gender, class, caste, wealth, age, ethnicity, religion, language, and other aspects divide and crosscut the communities. Beliefs, interests and values of community members may therefore conflict. Moreover, in stratified communities, local power relations play a significant role and, in the process, some are left powerless, while others enjoy social and political power. The nature of power relations is also determined by the increasing class segregation within communities. As Lewis (1999) remarks, disaster vulnerability comes not only from being poor, but also from being powerless to do anything about vulnerability, which results from actions and activities of richer or socially privileged, and therefore more powerful groups. Distribution of power is also connected to access to knowledge, since knowledge itself is power. Lack of education makes people powerless as they are not conscious of their strengths and also lose their ability to voice their concerns and demand their rights. Experts hold the power to make decisions for them. Such power structure within communities inhibits all good intentions of community participation for reducing disaster vulnerability.

However, achieving effective community engagement on ground and not as a rhetoric is indeed a challenge, especially in the light of changed social and economic context of heritage due to the urbanization process. This is even more challenging where there are already multiple layers of hierarchy, and ethnic and religious diversity within communities, as is the case in caste-based societies in countries like India.

Unfortunately, community participation approaches for disaster management in many instances are blind to the local power structure. The top-down approach among politicians and experts generally views the local community as a potential barrier to the implementation of techno-scientific solutions. It is assumed that what is needed is merely sufficient information to make the community accept those solutions and therefore community participation is essentially a case of enlightenment and not power and manipulation. In this way, community participation is made into a matter of disseminating scientific facts, intended to ensure acceptance and facilitate the implementation of any techno-scientific solutions. Such a perspective not only ignores the uncertainties of scientific knowledge, but also sidelines the community's experience-based knowledge, which is branded as "unscientific." In contrast to this view of the decision makers, the local community, based on its intimate knowledge of conditions in their own environment, may have major objections, apathy to, or ignorance of purely technical solutions.

Non-governmental organizations (NGOs) can also play a very important role in bridging the gap between government and local communities. This would necessitate effective awareness raising programs to sensitize community members on impending risks to lives and heritage and the significant role communities can play as volunteers in reducing risks as well as responding to disasters. This cannot be achieved if NGOs merely involve the communities in various disaster risk management activities. Real participation would entail gaining trust of the community through long term engagement. In Gujarat, NGOs that had been working in the affected region before the 2001 earthquake were more successful in their efforts compared to those who descended following the earthquake.

Moreover, the approach has to be that of a facilitator rather than a provider by helping communities take informed decisions. There are enough examples to demonstrate that patronizing approach can either hurt self-esteem of communities or make them dependent on external support. As already mentioned, it is also important to understand the internal dynamics, sensibilities and power structures within the community and sometimes approach various target groups such as those from lower social strata or women separately rather than the entire community as a whole. Otherwise the needs and constraints of weak and the marginalized may not be addressed or conflicts may surface among various sections.

7 Preconditions for achieving good governance

Effective mainstreaming of cultural heritage in disaster risk governance will necessitate some prerequisites that are briefly outlined below. First and foremost, it is important to achieve "policy coherence" so that different policies be from disaster risk management, climate change, development or cultural heritage management are able to speak to each other in a coherent manner. Having common grounds will be fundamental for achieving effective inter-sectoral coordination.

Another precondition is the establishment of communication channels that can help downscaling of international and national policies for implementation on ground, and at the same time successful grass-root community-based initiatives to be able to find resonance in the national policies. Both might adopt different means for the same end goals aimed at reducing disaster risks and effective response and recovery of cultural heritage following disasters.

Accountability and transparency in decision making are other important virtues of good disaster risk governance in general, and also in this case. This may be achieved by clearly delineating roles and responsibilities of the heritage sector at various levels to undertake various measures for reducing risks and responding during emergency situation in close coordination with disaster risk managers and civic defense agencies. This would also entail disaster risk governance that establishes clearly defined protocols for heritage and disaster management organizations; staff and communities linked to cultural heritage sites to effectively respond before, during and after disaster situations and periodically review and update these.

Last but not the least, equity and empowerment are two essential pre-conditions for good governance through community participation in disaster risk management for cultural heritage. Empowerment implies that the local communities are not just involved in disaster risk reduction and post disaster reconstruction efforts, but have a greater say and role in decision-making through genuine dialogue rather than one-way communication. This requires strengthening of internal communication systems within communities and also communication between community and the decision makers or "experts" and the politicians.

Giving "power to people" also implies that equity issues need to be resolved among the communities, and equity is not to be seen merely from a Western perspective. Rather, equity issues are very much dependent on locally defined roles and responsibilities and their acceptance through human dignity and support mechanisms. Equity of values defined from "outside" cannot be forced upon these communities. Since knowledge is power, "power to people" also implies widening and deepening the knowledge base and breaking the monopoly and ownership of knowledge. This would necessitate finding out ways of intervening with the local power structure, so that no one, irrespective of social status, economic class and gender is left out in the process of disaster risk reduction.

References

Benton, T., 2020. Coronavirus: why are we catching more diseases from animals? BBC. Available from: https://www.bbc.com/news/health-51237225.

Booker, F., Franks, P., 2019. Governance Assessment for Protected and Conserved Areas (GAPA). Methodology Manual for GAPA Facilitators. IIED, London.

Borrini-Feyerabend, G., Hill, R., 2015. Governance for the conservation of nature. In: Worboys, G.L., Lockwood, M., Kothari, A., Feary, S., Pulsford, I. (Eds.), Protected Area Governance and Management. Australian National University Press, Canberra, pp. 169–206. https://www.iccaconsortium.org/wp-content/uploads/2015/08/Governance-for-the-conservation-of-nature.pdf.

Borrini-Feyerabend, G., et al., 2013. Governance of Protected Areas: From Understanding to Action. Best Practice Protected Area Guidelines Series No. 20. IUCN, Switzerland.

Dean, M., Boccardi, G., 2015. Sendai implications for culture and heritage. Crisis Response J. 10 (4), 54–55.

Graham, J., Amos, B., Plumptre, T., 2003. Governance Principles for Protected Areas in the 21st Century, a Discussion Paper. Institute on Governance in collaboration with Parks Canada and Canadian International Development Agency, Canada.

Jigyasu, R., 2002. Reducing Disaster Vulnerability through Local Knowledge and Capacity. The Case of Earthquake Prone Rural Communities in India and Nepal (Unpublished Dr. Ing thesis). Norwegian University of Science and Technology, Trondheim.

Lewis, J., 1999. Development in Disaster-Prone Places—Studies of Vulnerability. Practical Action, Warwickshire, UK.

Seneviratne, S.I., Nicholls, N., Easterling, D., Goodess, C.M., Kanae, S., Kossin, J., Zhang, X., 2012. Changes in climate extremes and their impacts on the natural physical environment. In: Field, C.B., Barros, V., Stocker, T.F., Qin, D., Dokken, D.J., Ebi, K.L., Midgley, P.M. (Eds.), Managing the Risks of Extreme Events and Disasters to Advance Climate Change Adaptation. Cambridge University Press, Cambridge, pp. 109–230. A Special Report of Working Groups I and II of the Intergovernmental Panel on Climate Change (IPCC) https://www.ipcc.ch/site/assets/uploads/2018/03/SREX-Chap3_FINAL-1.pdf.

UNISDR, 2008. Links between Disaster Risk Reduction, Development and Climate Change. United Nations Office for Disaster Risk Reduction, Geneva. https://www.preventionweb.net/files/8383_pbdisasterriskreduction1.pdf.

UNISDR, 2013. Heritage and Resilience. United Nations Office for Disaster Risk Reduction, Geneva. https://www.undrr.org/publication/heritage-and-resilience-issues-and-opportunities-reducing-disaster-risks.

Chapter 4

Role of higher education institutions toward effective multi-hazard early warnings in Asia

Kinkini Hemachandra, Richard Haigh, and Dilanthi Amaratunga
University of Huddersfield, Huddersfield, United Kingdom

1 Introduction

The number of people living in low-elevation coastal zones is expected to increase from 58% in 2000 to 71% by 2050 (Merkens et al., 2016). These communities have been frequently affected by multiple hazards (Setyono and Yuniartanti, 2016) ranging from slow onset; coastal erosion and sea level rise, to rapid onset hazards; tsunami, oil spills, wind storms, and flooding. Further to this population expansion in coastal zones (Seto et al., 2011), changes in climatic conditions (Spalding et al., 2014) and other systematic risk factors for example, migration (Hugo, 2011) and socio-economic conditions (Spalding et al., 2014) make further complexities when dealing with coastal hazards.

Coastal hazards in Asia have become frequent and destroyed lives and economies. For example, the Indian Ocean Tsunami in December 2004 killed more than 230,000 people in 14 countries with the highest number of deaths reported from Indonesia, Thailand, Sri Lanka and India. Another example is, Cyclone Giri which destroyed more than 75% houses in Rakhine state in Myanmar in 2010 (Dutta and Basnayake, 2018). Typhoon Haiyan hit the Philippines on November 8, 2013 killing more than 6300 people, injuring more than 28,000 and affecting more than 16,078,000 (NDRRMC, 2013). Recent tsunami and earthquake incidents in Indonesia killed more than 2000 people in 2018 (CNN, 2018a). Of the reported 90 storms in 2015, 43 storms hit Asia and the Pacific region, with a massive impact on lives and economies (Dutta and Basnayake, 2018).

Rising levels of disaster risk in coastal regions have led to direct global initiatives aimed at introducing comprehensive disaster risk reduction (DRR) and resilience mechanisms. Multi hazard early warning (MHEW) is one such mechanism to reduce disaster risk and enhance resilience. The Sendai Framework for Disaster Risk Reduction (SFDRR) stresses the role of MHEW toward DRR. Its seventh goal aims to substantially increase the availability and access to MHEW and disaster risk information and assessment to people by 2030. The framework further highlights the necessity of effective regional cooperation in line with the Global Framework for Climate Services, to enhance sharing of information across countries (UNISDR, 2015). Effective MHEW comprises with number of enablers. However, evidence from Asia shows that existing early warning systems face many challenges in the region (Seng, 2013; UN-ESCAP, 2015).

In this effort, higher education institutions have been recognized as agents which facilitate in setting and operating effective MHEW. Their contribution is identified as a driver for assisting many enablers of an effective MHEW. Hence, the chapter presents the role of HEIs along with regional cooperation for enhancing coastal resilience through effective MHEW in Asia. This begins with Section 2 presenting coastal hazards and early warning systems in Asia.

2 Coastal hazards and early warning systems

2.1 Impact of coastal hazards

As explained in the introduction, coastal hazards have become frequent irrespective of economic and social development status of countries across the world (Harley et al., 2016). There are man-made causes and natural causes for coastal hazards. Coastal pollution, sea-freight transportation, industrial accidents and marine related infectious diseases are some sources of man-made coastal hazards whereas weather-related hazards (hydro-meteorological) and geophysical hazards are considered as natural causes of coastal hazards (Zou and Wei, 2010). Tropical cyclones, storms, floods and climate generated

Strengthening Disaster Risk Governance to Manage Disaster Risk. https://doi.org/10.1016/B978-0-12-818750-0.00004-0

TABLE 4.1 Impact of coastal hazards during 1900–2018.

Continent	Number of tsunamis	Number of deaths	Number of storms	Number of deaths	Number of cyclones	Number of deaths
Africa	4	312	83	2638	130	4484
America	10	1116	433	9233	665	88,946
Asia	36	266,266	304	9488	1196	1,253,085
Europe	4	2376	274	1984	23	203
Oceania	12	2810	–	–	236	1849

Source: EM-DAT data base 2018.

hazards are the most common weather-related natural hazards whereas earthquakes, tidal waves and tsunamis are some of the most common geophysical natural coastal hazards.

Both man-made and naturally induced coastal hazards have caused devastating impacts on human lives (Davis et al., 2015; Zou and Wei, 2010). During 1998–2017 more than 1.3 million fatalities were reported. 57% of reported fatalities were from tsunami related earthquakes and 18% were from storms related hazards during this period. 2004 Indian Ocean Tsunami is a landmark in coastal hazard history (Kottegoda, 2011). There were many other instances where coastal hazards have affected human lives.

For example, in 2005, the Hurricane Katrina incident in the United States caused 971 fatalities only in Louisiana state (Brunkard et al., 2005). Cyclone Sidr in Bangladesh killed 3406 people in 2007 (Paul, 2009). Similarly, Cyclone Nargis in 2008 caused more than 138,000 fatalities, more than 2.4 million affected people and more than one million homeless people in Myanmar (Besset et al., 2017). In Europe, the West Coast of France was hit by Xynthia Storm in 2010 causing 47 fatalities (Chadenas et al., 2014). Another major coastal hazard reported from the Pacific Region is the 2011 tsunami incident in Tohoku, Japan, with a death toll of more than 20,000 people (Nakahara and Ichikawa, 2013).

Among these reported coastal hazards, tropical cyclones (Mori and Takemi, 2016) and tsunamis (Saito, 2012) caused catastrophic impacts. Table 4.1 shows the impact of tsunamis, storms and cyclones in Asia when compared to other regions during 1900–2018. A similar finding is presented in a study by Zou and Wei (2010). According to them, Bangladesh, India, Indonesia, Malaysia, Philippines, Sri Lanka, Thailand, and Vietnam accounted for 29.1% of the total number of deaths and 56.67% of the total number of affected people by coastal hazards during 1985–2006.

In addition, coastal hazards have severe impacts on economies and development activities. For example, storm-related world economic losses were estimated as 1300 billion US$ during 1998–2017 (CRED, 2018). As similar examples, the economic loss of Cyclone Nargis was estimated as more than 10 billion US$ (Besset et al., 2017) whereas the economic loss of the 2004 Indian Ocean Tsunami for Sri Lanka was estimated to be 1 billion US$.

These figures highlight the severity of coastal hazards on lives and economies across the world.

Many researchers furthermore emphasize the future risk of increasing coastal hazards due to many factors. For example, rising sea levels may affect coastal flooding (IPCC, 2014; Merkens et al., 2016), inundation, erosion and salination in coastal regions (Gornitz, 1991). Studies also found that global warming could affect frequency, intensity and track of cyclones/hurricanes/typhoons (Mori and Takemi, 2016). Rapid development activities taken place in coastal areas may further enhance complexities of future coastal hazards (Merkens et al., 2016; Mori and Takemi, 2016; Spalding et al., 2014). Many researches argue that rising population may contribute toward causing hazards to be more frequent and more complex (Besset et al., 2017; Merkens et al., 2016; Mori and Takemi, 2016; Zou and Wei, 2010). Furthermore, Asia and the Pacific region is more vulnerable to coastal hazards due to its geographical locations (Hemachandra et al., 2019; Zou and Wei, 2010) along with rapid population rise in the future (Mori and Takemi, 2016).

2.2 Early warning systems

To minimize devastating impact of coastal hazards, especially from storms, cyclone, tsunamis and tsunami related earthquakes, effective disaster risk reduction strategies are recommended by many researchers, practitioners and policy makers (Davis et al., 2015; Mori and Takemi, 2016). Strengthening disaster preparedness, response mechanisms and early warnings have been recommended as effective strategies of disaster risk reduction (Basher, 2006; CRED-UNISDR, 2016;

Golnaraghi, 2012; UNISDR, 2015). Among these strategies, early warnings (EW) are considered as effective strategy for preparedness, response planning and effective recovery measures.

According to UNISDR (2009), early warning systems are defined as "The set of capacities needed to generate and disseminate timely and meaningful warning information to enable individuals, communities and organizations threatened by a hazard to prepare and to act appropriately and in sufficient time to reduce the possibility of harm or loss." In its next revision in 2017, it is re-defined as "An integrated system of hazard monitoring, forecasting and prediction, disaster risk assessment, communication and preparedness activities systems and processes that enables individuals, communities, governments, businesses and others to take timely action to reduce disaster risks in advance of hazardous events" (UNISDR, 2017). These two definitions emphasize elements of an EW system and its objectives.

Further, the World Meteorological Organization (WMO) sets out key elements of an effective EW system as presented below.

- **Disaster risk knowledge**: systematic collection and analysis of data along with hazard dynamics and vulnerabilities
- **Detection, monitoring, analysis and forecasting of hazards and possible consequences**: continuous monitoring of hazard parameters for issuing accurate and timely warning
- **Warning dissemination and communication**: use of standards and protocols when issuing warnings to target groups
- **Preparedness and response capabilities**: through proper educational and awareness programs for people to inform on options for safe behavior to reduce disaster risks (WMO, 2018)

Coordination of those elements across many agencies from national to local level is necessary for its success. Different stakeholders are responsible for the elements in the EW system. For example, issuance of EW is a national responsibility while implementation of EW is a responsibility of public and private stakeholders. Further, for effective EW systems, supportive regulatory frameworks, plans, budgetary allocations and operational mechanisms are identified as essentials (WMO, 2016). In addition, state-of-art technology and accurate forecasting are similarly important for its success. However, advanced technologies and forecasting do not benefit communities at risk, if they are not informed in a timely manner. Similarly, well-prepared communities will not be saved if they do not have access to clear and reliable information (Jubach and Tokar, 2016). In summary, failure to achieve one of these components or lack of coordination across each component will result in failure of the whole system (UNISDR, 2017).

Moreover, effective EWs are considered to be end-to-end systems. Risk information, warning dissemination and response systems are interconnected from upstream, interface to downstream (Jubach and Tokar, 2016). Similarly, EWs are assumed to be people-centered. A people-cantered EWs has the ability to empower individuals and communities threatened by hazards and to act in sufficient time and in an appropriate manner to reduce possible personal injuries and illnesses, loss of lives and damage to properties, assets and the environment (WMO, 2018).

2.3 Importance of early warning systems

Empirical evidence shows how early warnings have benefitted society to reduce effects of hazards (Basher, 2006; Harley et al., 2016; Mori and Takemi, 2016; Paul, 2009). For example, the Government of Cuba issued timely EWs before Hurricane Charley hit in 2004. As a result, they were able to limit the number of reported fatalities to four even after severe destruction of more than 70,000 houses. Right after Hurricane Charley, Cuba was affected by Hurricane Ivan again in 2004. At this occasion too, the Cuban Government's EW system was able to evacuate 2 million people without a single fatality (UNISDR, 2005).

In another example, the EWs in Bangladesh helped to reduce the number of fatalities from Cyclone Sidr in 2007. It affected southwestern coast of Bangladesh and caused 3406 fatalities. Even though this is a significant number, when compared to Cyclone Gorky in 1991, this number is insignificant. Cyclone Gorky hit Bangladesh and killed more than 140,000 people in 1991. This significant level of reduction of loss of lives was identified as a result of timely forecasting and effective early warnings systems operated in Bangladesh (Paul, 2009).

2.4 Global initiatives

In addition, the importance of effective early warnings has been recognized at number of global level platforms highlighting their ability to save lives (CRED-UNISDR, 2016; UNISDR, 2005). For example, the first International Conference on Early Warning Systems was held in 1998 in Potsdam, German. In parallel, the International Strategy for Disaster Reduction was established by the UN General Assembly in 2000. It provided recommendation for a framework for advancing early warnings as a vital risk management tool for natural and technological hazards (UNISDR, 2003). This is furthermore emphasized at the Second International Conference on Early Warning held in 2003 in Bonn, Germany. With participation of 1250 experts, a checklist was developed as an outcome of the Third Early Warning Conference held in March 2006 in Bonn, Germany. The checklist helps to identify any gaps/challenges in the existing early warning systems across the world. This conference was called after the 2004 Indian Ocean Tsunami incident to address the urgent need of developing early warning systems worldwide (UNISDR, 2006).

In parallel, the Hyogo Framework for Action (HFA) 2005–2015 was introduced at the World Conference on Disaster Risk Reduction held in Hyogo, Japan in 2005. This framework aimed at building the resilience of nations and communities to disasters during 2005–2015. At the meeting, the necessity of integrating disaster risks into policies, plans and programs for sustainable development was acknowledged (UNISDR, 2005). The HFA was ratified by 168 signatories as a paradigm shift in disaster risk management from emergency response to a comprehensive risk reduction approach inclusive of preparedness and preventive strategies (WMO, 2016). Its second priority for action was to identify, assess and monitor disaster risks and enhance early warnings highlighting the special role of early warnings in risk reduction (UNISDR, 2005). HFA set the background for the importance of EW and hence encouraged development of EW as people-centered EW to ensure warnings are timely and understandably disseminated to people at risk.

2.5 Issues/challenges in early warning systems

Despite the importance of effective EW, there is evidence which recommends further improvements to be made for existing disaster risk reduction measures (Harley et al., 2016; Paul, 2009; Seng, 2013; UN-ESCAP, 2015). According to a study by Paul (2009), 22% of surveyed respondents revealed that they did not receive timely warnings before the Cyclone Sidr hit in Bangladesh. He identified reasons for this failure as weak operations among authorities and technical issues when disseminating warning messages. For example, the warnings were not heard by residents due to wind speed and its direction.

Another issue highlighted was under or over estimation of hazards risk within present EW (CNN, 2018b; Harley et al., 2016). For example, when Halloween Storm hit in Italy in 2012, the authorities were unable to forecast for six forecast sites with high hazard code-red. That is due to underestimation of extreme water levels in the forecasts sites. That led to severe damage in the areas by coastal flooding.

Another challenge identified within existing EW systems is their single-hazard focus. There are many instances where multiple hazards are recorded. For example, recent incidents in Indonesia show multiple hazards, including tsunamis, along with soil liquefaction and earthquakes. In Japan in 2011, the Great East Japan Earthquake triggered a massive tsunami, similar to the situation with the 2004 Indian Ocean tsunami. In 2016 floods and landslides were reported in Sri Lanka. To address these multiple hazards, experts proposed EWs are required to be multi-hazard focussed.

3 Multi-hazard approach

3.1 Introducing multi-hazard early warning (MHEW)

This situation requires researchers, practitioners and policy makers to introduce and strengthen MHEW systems to reduce future disaster risks and enhance resilience. United Nations International Strategy for Disaster Reduction (UNISDR) defines the term multi hazard as "The selection of multiple major hazards that the country faces, and the specific contexts where hazardous events may occur simultaneously, cascading or cumulatively over time, and taking into account the potential interrelated effects" (UNISDR, 2017).

Further, the report underlines the role of MHEW system for addressing several hazards and/or impacts of similar or different type in contexts where hazardous events may occur alone, simultaneously, cascadingly or cumulatively over time, and considering the potential interrelated effects. A MHEW system with the ability to warn of one or more hazards increases the efficiency and consistency of warnings through coordinated and compatible mechanisms and capacities. This involves multiple disciplines for updating and accurate hazard identification and monitoring for multiple hazards (UNISDR, 2017).

3.2 Global initiatives on multi-hazard early warning

At first, the concept of MHEW was initiated at the Third Early Warning Conference held in 2006. Participants at the conference demanded an integrated and holistic approach to early warnings for multiple hazards and risks (WMO, 2018) due to increasing globalization of risks of natural and manmade hazards (Alfieri et al., 2012; Golnaraghi, 2012; WMO, 2018).

Later, in 2015, three similar initiatives have highlighted the necessity of MHEW at various global platforms. First, the Sendai Framework for Disaster Risk Reduction highlighted the significance of MHEW in 2015. The framework guides disaster risk reduction strategies at global, regional and national level. It identified seven global targets with four priority actions. Its seventh target aims to substantially increase availability of and access to MHEW and disaster risk information and assessments to people by 2030 (CRED and UNISDR, 2016).

The second global initiative which promotes the concept of MHEW is the Paris Agreement. Its Article 7, Section 7(C) highlights the necessity of strengthening scientific knowledge on climate system and early warning systems for better decision making (UNFCCC, 2015). A third major global initiative introduced in 2015 was the Sustainable Development Goals (SDGs). It recognizes the significance of global partnerships enabling poor and vulnerable countries, through provisioning of resources for effective risk reduction and resilience. Specifically, its 13th Goal emphasizes the importance of enhancing human and institutional capacities in MHEW (UN, 2015). In addition, EWs are considered as an important element in European Civil Protection Mechanism as a risk prevention tool (Alfieri et al., 2012).

Apart from these initiatives, the first Multi-Hazard Early Warning Conference (2017) held in Cancun, Mexico in 2017 called to learn and exchange information and to promote adaptation of good practices in individual, cluster and MHEW. The conference further aimed to find out how countries can improve accessibility of and access to MHEW with risk information and assessments. At this conference, a checklist was presented as a conference outcome. The conference further presented key elements of an end-to-end as well as people-centred multi-hazard-early warning system.

In addition to these global initiatives, many international, regional and national levels initiatives have been taken place on multi hazard forecasting and early warnings. For example, the European Flood Alert system provides international collaboration with national authorities for addressing trans-boundary flood hazards in Europe. Similarly, the French rainfall flood vigilance system is an example of MHEW in France at national level. There, Meteo-France provides meteorological vigilance maps twice a day regarding different types of extreme weather including coastal hazards in France (Alfieri et al., 2012).

In addition to these policy initiatives, many researchers emphasize the importance of multi-hazard approaches as effective strategies for disaster risk reduction. They emphasize their ability to achieve economies of scale, sustainability and efficiency of EW (Rogers and Tsirkunov, 2011). Further, researchers identify the ability to enhance preparedness measures among communities through better understanding of the multiple risks they face (Rogers and Tsirkunov, 2011) due to increasing frequency and magnitude of hazards (Alfieri et al., 2012; Rogers and Tsirkunov, 2011). Hence, the issue raised here is to identify enablers for an effective MHEW which is explained in the next section.

3.3 Enablers of effective multi-hazard early warning systems

To develop and operate an effective MHEW specifically for coastal resilience, the following factors have been identified as enablers, based on literature review. These enablers were categorized into three categories as presented in Fig. 4.1 based on their similarity among variables:

(1) Policy, legislative and institutional arrangements
(2) Social and cultural considerations
(3) Technological and scientific arrangements

Policy, legislative and institutional arrangements

Governance

Governance is considered as a fundamental element of an effective DRR strategy (Gall et al., 2014), and has the capacity to enhance effectiveness, accountability and transparency of risk reduction strategies (Hemachandra et al., 2018; Renn, 2012; Rogers and Tsirkunov, 2011). Governance is significantly appropriate for developing countries which face difficulties or capacity gaps within administrative, organizational, financial and political boundaries when dealing with hazards (Ishiwatari, 2012). Risk governance is particularly important when making collective decisions under any uncertain situations (Renn, 2012). One of the key elements of risk governance is multi-stakeholder participation in risk reduction decision making. Among these stakeholders, governments play a major role as the sovereignty holder engages in all legal and administrative matters and has the power of assigning and allocating necessary recourses (Meerpoël, 2015).

Researchers emphasize that well-developed governance structures along with robust legal and regulatory frameworks, political commitment and effective institutional arrangements help the development and sustainability of sound early warning systems (Harley et al., 2016; Rogers and Tsirkunov, 2011; Seng, 2013). Moreover, governance facilitates resource allocation as well as better coordination between science and policy (Basher, 2006; Meerpoël, 2015).

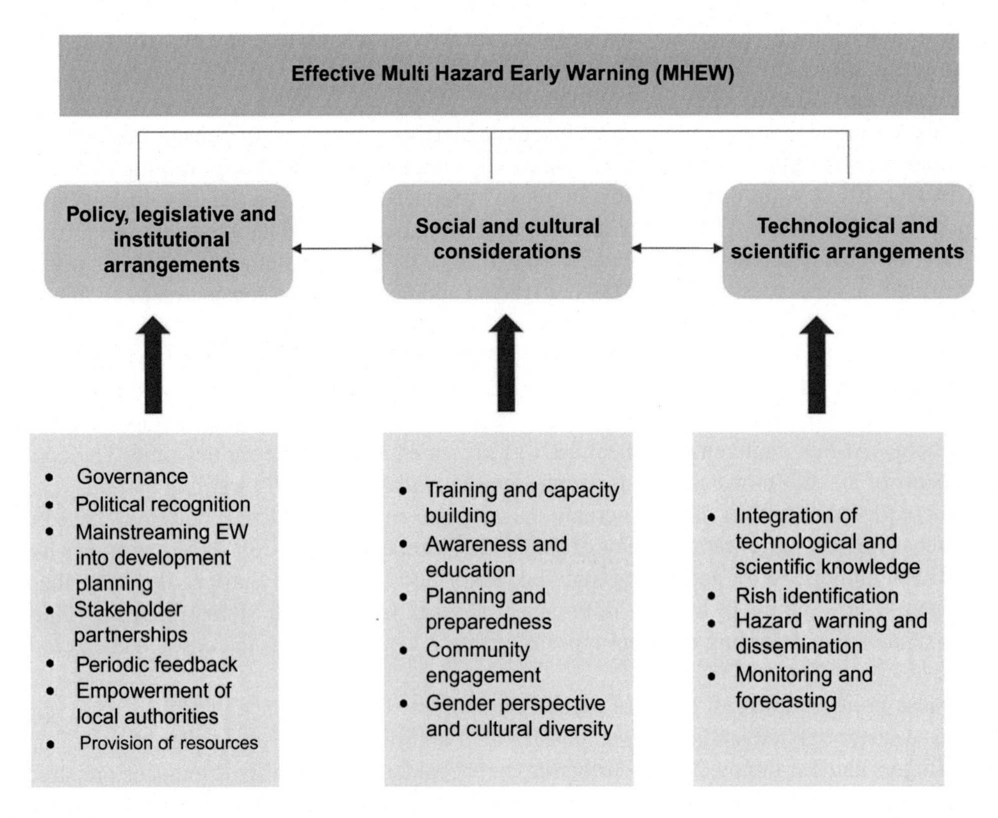

FIG. 4.1 Enablers of effective multi-hazard early warning systems.

Political recognition

Political recognition or political support affects the success of any DRR strategy. Sound political recognition has the ability to determine priorities, has the power to introduce new legal provisions as well as to influence resource allocation (Rogers and Tsirkunov, 2011). Strong political recognition about benefits of EWS are reflected in harmonized national to local disaster risk management policies, planning, legislation and budgeting (Golnaraghi, 2012). Most EW systems are established and run by states. This shows the importance of political sponsorship for implementing and operating EW systems. However, when an EW system delivers results which are against the government policies and agendas, the level of political support may reduce (Kelman and Glantz, 2014).

A case of the Philippines explains how the political support and acceptance saved lives during Typhoon Koppu in 2015. The Government of the Philippines introduced a Zero Casualty policy after following the Typhoon Haiyan in 2013. This new policy lead toward substantial reduction in terms of loss of lives and number of affected people from Typhoon Koppu (UN-ESCAP, 2015).

Another example can be found in Bangladesh. After super cyclone in 1970, the Government of Bangladesh and the Red Cross together established Cyclone Preparedness Programme. In addition, the government has introduced several legal provisions to support their DRR strategy (Habib et al., 2012).

Mainstreaming EW into development planning

State-of-the-art technological and scientific knowledge do not ensure effective EWs as risk reduction strategies (UN-ESCAP, 2015). EW are required to be incorporated into development planning for its success and sustainability (UNISDR, 2015). Many researchers claim that properly integrated EW to institutional mechanisms has the ability to reduce disaster risks and enhance resilience (Alfieri et al., 2012; Zia and Wagner, 2015).

At the global level, the Sendai Framework for Disaster Risk Reduction emphasizes the necessity of integration of MHEW into development planning process across developed and developing nations (Zia and Wagner, 2015). For a successful integration of EW into development planning, resource allocation and changes to the governance structure are further recommended.

For example, in the African region, EW on meteorological changes have been incorporated to land-use planning and food planning policy in Africa to address food security issues in the region (Challinor et al., 2007). This can be facilitated by establishing key institutions at national and local level (Alfieri et al., 2012).

Stakeholder partnerships

Another important element for an effective MHEW is stakeholder partnerships. No single institute is able to provide a fully comprehensive solution toward DRR. Hence, researchers promote stakeholder partnerships for successful implementation of EW and other DRR strategies (Jubach and Tokar, 2016; Rogers and Tsirkunov, 2011).

Jubach and Tokar (2016) outlined the role of stakeholder partnerships at national and regional level when addressing transboundary hazards. These partnerships provide opportunities to share expertise from a broader field of knowledge, better use of limited financial resources at national level and sharing expensive EW systems regionally, with lessons learned. These partnerships collaborate expertise from a wide range of disciplines and provide consistent messages from multiple sources (Jubach and Tokar, 2016; Rogers and Tsirkunov, 2011).

Legislations, directives, Memorandums of Understanding (MoUs) and institutional arrangements help in determining clear roles and responsibilities among different organizations as well as in facilitating their existence and operation (Golnaraghi, 2012; Rogers and Tsirkunov, 2011). Stakeholder partnerships can be developed among government agencies, media, non-government organizations, academic institutions and schools, emergency relief and humanitarian agencies, meteorological societies, private sectors and utility service providers (Rogers and Tsirkunov, 2011).

Periodic feedback

Effective feedback mechanisms are considered as essential for improvement of overall EW. Periodic feedback provides inputs for planning, coordination, operational and technical aspects of the EWS (Golnaraghi, 2012). User feedback also helps to identify weaknesses or limitations in the existing system and accordingly, make recommendations (Jubach and Tokar, 2016).

For example, the Government of Japan has established a Technical Investigation Group on Inheritance of Lessons from Disasters (TIGLD) to review EW in Japan to identify problems and challenges experienced in disasters and drawing suggestions for further improvements (Hasegawa et al., 2012).

United States, another example, has implemented National Weather Services as a strategy of collecting feedback (NWS). Local NWS offices are responsible to evaluate and review operations after a disaster event to identify best practices as well as lessons learned. Their objective is to further improve existing EW. They use local emergency management feedback and media as source of feedback. They further conduct formal assessments through emergency management communities and NWS peers (Buan and Diamond, 2012).

The Government of Bangladesh similarly provides feedback on their existing EWS as a part of Standing Order on Disasters (SOD). They collect feedback from the government, media, volunteers, and general public. Their SOD assign responsibility to each person to record any weaknesses in the system as well as to make recommendations (Habib et al., 2012). Hence, effective feedback and improvement mechanisms are recommended to be placed at all levels of EW to provide systematic evaluation and ensure system improvement over time (Golnaraghi, 2012).

Empowerment of local authorities

Most hazards are local in nature. Hence, DRR strategies can be initiated and implemented at local level institutions. Local governments or local authorities are aware of local hazards, their significance, local communities and their capacities (Scott and Few, 2016). Local authorities are very important in EW dissemination and acting on EW alerts. EW received from national agencies need to be communicated to communities at risks through local governments in time and reliable manner. There, local authorities can play a significant role in this effort as the lower level of state representation in many countries. They are considered as the closest government unit to citizens.

A public opinion survey conducted by Pew Research Center for the People and the Press (2013) revealed that communities are more favorable for local government than federal government in the United States. Hence, they believe disaster information given by local governments than other government agencies. A similar finding is revealed by the study of Collins (2009a); local government's EW radio alert system can play a major role by communicating EW through radio system for alerts on hazards. He further explains that local government can also give feedback to the warning providers regarding warning messages to identify weaknesses or challenges.

Provision of resources/infrastructure

Success of any risk reduction program depends on availability of resources and infrastructure. Provision of adequate human, financial and other resources is necessary for their successful operation and sustainability (Golnaraghi, 2012). Most early warning systems are funded and maintained by governments (WMO, 2016) whose funding availability is limited to their budgetary capacities.

For example, the Government of Japan finance EW and other DRR programs through national and local governments' budgets. Their allocations have been further increased after a disaster to restore and future disaster prevention efforts (Hasegawa et al., 2012). In addition, governments allocate resources such as aeroplanes, boats and other assets of armed forces and coast guards as precautionary efforts (UN-ESCAP, 2015). For example, in Hong Kong, early warnings are supported with other infrastructure facilities such as transportation and communication network during a disaster to help communities affected in hazards (Rogers and Tsirkunov, 2011). However, budgetary share allocated for preparedness measures including early warning systems are comparatively low as compared to post disaster recovery and relief efforts (Healy and Malhotra, 2009; Seng, 2013). Researchers explained this is due to inability to estimate benefits of prevention of disasters (Rogers and Tsirkunov, 2011).

In addition to the above-mentioned policy, legislative and institutional factors, social and cultural factors have been identified as enablers for effective MHEW.

Social and cultural considerations

Training and capacity building

Training and capacity building are central for effective functioning of EW since they provide necessary education and awareness for operation of EW according to predetermined plan (Alfieri et al., 2012). Community training can enhance community disaster preparedness culture to act upon EW alerts (Chaimanee, 2006). Similarly, training for institutional members can enhance integration of EW into planning and policy making at all levels of governments (Alfieri et al., 2012).

For example, the Government of Tamil Nadu with the assistance of the UNDP, undertook a project to strengthen and institutionalize EW within coastal districts affected by tsunamis in Tamil Nadu. The program trained more than 1500 people and gave a training manual published in both English and Tamil languages. That training focussed on raising awareness and building capacities among communities in coastal risks in Tamil Nadu in India (Chaimanee, 2006).

Training can be incorporated into formal and informal educational programs. Informal trainings are provided through talks, conferences, radio and TV channels whereas formal trainings are provided through short courses, school curricular, and producing training manuals (Golnaraghi, 2012; WMO, 2016). In addition, regular drills and tests can also be conducted among stakeholders who engage in emergency procedures (Alfieri et al., 2012). Training on EW systems covers many aspects. For example, raising awareness, identifying potential risks, recognizing warning signals during an event and accordingly to take necessary actions (Rogers and Tsirkunov, 2011).

Awareness/education

Public awareness is another vital enabler for designing effective EW system and its implementation (Alfieri et al., 2012; Paul, 2009; Rogers and Tsirkunov, 2011). The importance of community awareness on natural hazards was suddenly raised after 2004 Indian Ocean Tsunami incident (Basher, 2006). Better educated communities are considered as more resilient than others in the society (Frankenberg et al., 2013).

For example, the Government of Bangladesh has created awareness among coastal communities to act on early warning messages prior to Cyclone Sidr. As a result, they were able to reduce the number of deaths from the disaster (Paul, 2009). A similar finding is reported by Frankenberg et al. (2013) from the Indonesian context. They found that better informed and educated Indonesian males survived during 2004 Indian Ocean Tsunami when compared to women. Similarly, educated communities were found to be doing well in the evacuation centers compared to others. The study further highlighted that educated communities reported better health conditions after the disasters.

Hence, communities should be aware of the threats generated by different hazards, their potential impacts, threat levels indicated by warning messages and recommended actions (Golnaraghi, 2012). Ongoing education and awareness can be implemented via schools and universities within their curricular systems in addition to use of mass media. Further it has to be tailored to meet different needs of different communities (Rogers and Tsirkunov, 2011). Awareness can be introduced through formal and informal education at different levels in the society. Formal education can be introduced through introduction of modules and courses to school and university curricula whereas informal education and awareness can be introduced through training, drills, and through mass media and social media.

According to Tang et al. (2012) in China, education and awareness on MHEW to local communities were given through internet, media, brochures on MHEW, posters, training and exercise. Another special feature of Shanghai MHEW is that public awareness is carried out through special channels which broadcast weather-related programs. In addition, an image education program was conducted among 60 primary schools with more than 50,000 students in Shanghai. This program was aimed to raise awareness on meteorological hazards among students. For further improvement, existing education and awareness programs are evaluated to identify their weaknesses (Rogers and Tsirkunov, 2011).

Planning and preparedness

Disaster preparedness and response planning are other key factors for reducing disaster effects in any society (Alfieri et al., 2012). There is ample evidence to suggest that well-planned DRR strategies have reduced loss of lives and economic losses in global disaster history.

For example, the preparedness program on evacuation routes, shelters and EW communication methods in Bangladesh was able to reduce all types of losses from Cyclone Sidr when compared to Cyclone Gorky (Paul, 2009). They also reduced post disaster response and recovery costs (UNISDR, 2015).

Furthermore, it is accepted to support preparedness and response planning with scientific and technical information. For example, the Japanese Basic Disaster Management Plan is supported with scientific information for assessing disaster risks for tsunamis, earthquakes and other types of disasters (Hasegawa et al., 2012). Another example of a different type of preparedness measures can be identified from Shanghai's MGHEW system. Further, these plans can be integrated with governments' development plans (Alfieri et al., 2012). These plans and preparedness measures should not be ad-hoc, fragmented and lacking proper coordination with other development plans. Emergency response plans are developed with consideration for hazard/risk levels and characteristics and needs of exposed communities as well as other stakeholders such as emergency respondents, hospitals, schools, and campgrounds (Golnaraghi, 2012).

Community engagement and empowerment

One of the key stakeholders of people-centered EW is community (Rogers and Tsirkunov, 2011). Community has been recognized as the target group of any last mile warning system (UNISDR, 2015). Thomalla and Larsen (2010) emphasize that community representation in EW design is considered as a bottom-up approach. Their absence may hinder the capacity of hazard responses (Scott and Few, 2016).

According to (Collins, 2009b) a well-informed and alerted community enhances the effectiveness of EW system. When communities are aware of risk priorities, they are motivated to respond to early warning systems without waiting for warnings from outsiders (Thomalla and Larsen, 2010). Similarly, raising community awareness helps in creating a preparedness culture in the community (Kadel, 2011). Community awareness can be given on available safe options, escape routes, and mechanisms to avoid and minimize life and property damages (Rogers and Tsirkunov, 2011).

According to Rogers and Tsirkunov (2011), traditional and indigenous knowledge can be introduced to enhance effectiveness of EW. One such successful community engagement program was conducted by the Government of Bangladesh when disseminating EW messages to communities at risk. More than 42,000 community members were engaged in disseminating EW messages using hand sirens and megaphones. They disseminated warning messages released by the Bangladesh Meteorological Department through its cyclone Preparedness Program (Habib et al., 2012).

One of the weaknesses among communities is their poor level of responses to most hazard warnings due to many reasons. For example, Paul (2009) mentioned that during the Cyclone Sidhr, some community members did not follow evacuation orders and either stayed at their homes or returned to their homes from safe locations. Lack of adequate evacuation shelters in their area, inadequate spaces in those shelters and the distance from their homes to evacuation centers were highlighted as reasons for not following EW orders. According to Paul, reported fatalities and losses can be further minimized if communities follow these orders. Hence, further actions need to be taken to enhance community participation in MHEEW development.

Gender perspective and cultural diversity

Different hazards have different level of impacts for diverse communities (Islam et al., 2017). Hence, identification of their problems and needs during an event is essential in developing a successful EW. Specially, cultural differences and gender are important to be considered since these factors can influence their preparedness and response capacities greatly (Rogers and Tsirkunov, 2011). Specially, men and women have different roles in the society. Hence, communication of warnings could be different for different target groups during a disaster. Specially, women have limited access to EW information (Collins, 2009b). For example, most women stay at home during daytime whereas men have gone outside for work. Accordingly, these gender perspective should be considered since men and women engage in different social roles and are bound with different social and cultural norms (Rogers and Tsirkunov, 2011).

In addition, socio-economically vulnerable groups such as elderly, disabled communities are similarly considered when developing MHEW. According to Mayhorn and McLaughlin (2014), technological aspects of warning communication has been considered whereas cultural diversity and community characteristics have not been considered in most situations. Hence, they propose to utilize more systematic approaches when designing warning messages considering end users' cultural and personal values and believes.

Technological arrangements

Integration of technical and scientific knowledge

Technical and scientific information are considered as the third pillar of enablers of MHEW. They provide basis for risk identification and risk assessment in an effective EW system (Lauterjung et al., 2010; WMO, 2016). Sound knowledge of meteorological or hydrological phenomenon provide basis for predictions, preparedness and response planning measures (Rogers and Tsirkunov, 2011). Hence, their findings are necessary to be incorporated to institutions' operating systems (Zia and Wagner, 2015).

For example, the state of Virginia used Coastal GeoFIRM for the development of Digital Flood Insurance Rate Map at the Federal Emergency Management Agency of the United States to assess flood risk and to make flood recovery maps. The maps are furthermore tailored to match each state's specific spatial considerations (Valenzuela and Gangai, 2008). Similarly, Indonesia uses the Tsunami Early Warning System to identify tsunami risks in the Indian Ocean after the 2004 Indian Ocean Tsunami incident. They need the most accurate scientific information to identify and assess tsunami risks than other countries since they have very limited time of issuing EW and taking actions. Hence, The Government of Germany also provides further technical and scientific support for the development of the Tsunami Early Warning system in Indonesia for more reliable and accurate tsunami EW system (Lauterjung et al., 2010; Strunz et al., 2011).

Risk information

Another vital element of an effective MHEW is risk information. Hydro meteorological data, climatological data and some socio-economic and demographic data are considered as risk information. They are widely used for vulnerability assessments (WMO, 2017). Risk information provides hazard, exposure and vulnerability information to communities as well officials. They are useful when designing emergency planning, drawing risk maps and developing warning messages to communities and authorities (Golnaraghi, 2012; Strunz et al., 2011).

Risk information is also used for sustainability of EW systems (Basher, 2006; Thomalla and Larsen, 2010). Daily briefings, bulletins, special reports, websites and workstations are used in the process of formulation and dissemination of risk data. For example, the Japan Meteorological Agency provides risk information to public and municipal authorities through their websites and publications to deliver risk information in a timely an precise manner to interested groups (Hasegawa et al., 2012). According to (Rogers and Tsirkunov, 2011), risk information should be tailored to specific needs of users.

In addition, sharing risk information within a region helps to maintain cost effective early warning systems (UN-ESCAP, 2015). For example, the Indian Ocean Tsunami Early Warning Systems (IOTWMS) shares risk information and deals with issuing warning messages to its member countries (UNESCO-ICO; UN/ISDR/PPEW; WMO, 2005). This enhances member countries' access to reliable data through regional data bases. This makes it possible to issue timely warning messages to people at risk efficiently (Rogers and Tsirkunov, 2011).

Hazard warning dissemination and communication

Hazard warning dissemination and communication is another essential aspect in people-centered, end-to-end early warning systems (WMO, 2016). Warnings are expected to be disseminated and communicated to people who are at risk in timely, clearly and reliable manners to take required actions (Collins, 2009b; Golnaraghi, 2012). Warnings are mostly issued by state agencies, for example, meteorological offices, disaster management centers and ministries (UN-ESCAP, 2015; WMO, 2016). They use different modes of communication to issue hazard warning: faxes, SMS, emails, telephones, the internet, radio, colored flags and so on (WMO, 2016).

It is also important to assure that hazard warnings are issued with more specific information. According to Alfieri et al. (2012), location, hazard type, expected onset and its duration, possible damages and communities to be affected are some of the specific information to be included in warning messages. This will enhance reliability of hazard warnings among communities.

Similarly, hazard warnings should be clear. Clear hazard warnings can be issued using pre-determined warning mechanisms. Simple, useful and usable messages should be issued when disseminating and communicating hazard warnings. Many countries use color codes, numbers or hazard levels when communicating and disseminating warning messages (Golnaraghi, 2012).

Vertical and horizontal communication of EW among stakeholders is also essential for effective EW (Rogers and Tsirkunov, 2011). Furthermore, it is necessary to ensure that communication of information to emergency management authorities provides adequate lead time for initiating preparedness actions (Rogers and Tsirkunov, 2011). Even though vertical and horizontal communication is an important element in a well-functioning EWS, there is lack of attention to operational contexts (Thomalla and Larsen, 2010). Researchers emphasize that warning should be communicated with reliability. There are many instances where communities did not believe warnings and did not take actions accordingly. For example, according to (Alam and Rahman, 2014), disbelief of hazard warnings had greatly affected their evacuation decisions during the Bangladesh cyclone in 1991.

In addition, communication gaps between scientific community and policy makers (IFRC, 2009) and language barriers are some issues identified when communicating hazard warnings. For example, language differences among diverse communities has become a challenge to convey warning messages in Asia (UN-ESCAP, 2015). Another important consideration when disseminating hazard warnings is to ensure that dissemination methods can survive in any conditions like lack of electricity, wind speed, etc. According to Paul (2009), wind speed and wind direction affected warning efficiency during Bangladesh Cyclone in 1991. A similar incident was reported from Indonesia in 2018. Before the recent two tsunami incidents, hazard warnings were not disseminated to communities due to electricity failure (CNN, 2018b).

Monitoring and forecasting

Hazard monitoring and forecasting is one of the key elements in people cantered and end-to-end EW system (Rogers and Tsirkunov, 2011; WMO, 2017). This helps in identifying future risks through hydro meteorological parameters and thus provisioning of archived and real-time data, conducting hazard mapping and analysis and forecasting their patterns.

For example, in the United Kingdom, the United Kingdom Coastal Monitoring and Forecasting (UKCMF) network has been set up to monitor and forecast coastal hazards in the United Kingdom. This initiative works with forecasters, coastal authorities, academics, government, emergency respondents, industry and public. They provide prediction for tide levels and real-time observations of waves and tide data. They also forecast coastal flooding and accordingly advise relevant authorities for taking necessary actions using their archived data and information. The network work with European partners to share knowledge, data, research and forecasting techniques. Their main aim is to provide coastal flood risk forecasts and to provide long term evidence for developing long term forecasts models.

The factors presented above are identified based on literature as the critical enablers for strengthening MHEW toward coastal resilience. However, there are many challenges or gaps identified when developing and implementing MHEW. For example, there are governance issues (Seng, 2013), community engagement issues (Paul, 2009), hazard warning and dissemination issues (Thomalla and Larsen, 2010), limited capacity building and training opportunities and so on in the MHEW. To address these challenges or gaps require more resources and capacities which are expensive as well require more knowledge and expertise. Hence, higher education institutions are identified as key players who have the knowledge, capacity and expertise to contribute in the development and implementation of MHEW. Along with HEIs, a strong regional cooperation is further recommended by many researchers to deliver effective MHEW for coastal resilience. Hence, Section 4 presents the role of HEIs followed by Section 5 presenting regional cooperation to facilitate this effort.

4 Higher-education institutions (HEIs)

4.1 The role of higher-education institutions

Higher education institutions refer to traditional universities and profession-oriented institutions including colleges and further education institutions. Accordingly, HEIs offer and deliver higher education in many disciplines (IGI Global, 2019). The HEIs in the United Kingdom are considered to universities and or any institutions conducted by a higher education corporation, or any institution designated as eligible to receive support from funds administered by the Higher Education Funding Council for England (HEFCE) (Euroeducation, 2014). In the early days in the United Kingdom, HEIs were limited only to universities. However, within the last three decades, the scope has been expanded to include nonuniversities such as polytechnics, other institutions, and colleges of higher education (Kogan and Becher, 1980).

Their primary role is to raise awareness, skills, knowledge and values of societies for a sustainable future through generating professionals in specialized areas (Cortese, 2003). In addition, they engage in research and innovation activities which are critical inputs to growth in knowledge based assets (World Economic Forum, 2017). In addition, HEIs provide the basis for economic and social development of countries (Hatakenaka, 2015) through training and capacity building with broad range of disciplines and expertise (World Economic Forum, 2017). Furthermore, HEIs prepare most professionals who develop, lead, manage, teach, work in, and influence society's institutions for the betterment of the society (Cortese, 2003).

Research is a core function of HEIs which leads to enhance productivity and innovation (Liyanage et al., 2018), through developing new knowledge and ideas, commenting existing systems and engage in researching activities (Cortese, 2003). According to Seidel (1991), HEIs serve society by providing education and training as a combination of research and teaching. They also emphasized the role of HEIs in regional development and fostering intellectual and social development of society.

4.2 Role of higher-education institutions in disaster risk reduction and resilience

The role of HEIs in the present disaster context is emphasized by many researchers. Their primary role in disaster risk reduction is to *provide necessary skills and knowledge* for public and policy makers for making policies and plans related to DRR (Koehn, 2013; Perdikou et al., 2014). Further, HEIs play a vital role in *capacity building* among stakeholders who engage in DRR. They also help *in knowledge creation and dissemination of disaster resilience through* developing curricula or through research work (Hemachandra et al., 2019).

The SFDRR encourages the bringing together of *multiple stakeholders to produce disaster risk reduction through science and technology* (Aitsi-Selmi et al., 2016) while emphasizing their specific role toward disaster resilient society (UNISDR, 2015). Before the SFDRR, HFA also highlighted the need of disaster related education (UNISDR, 2005). This has been evident with the participation of more than 50 scientists in making 2013 Global Assessment Report and more than 200 representatives making the Special Report on Extreme Events by the Intergovernmental Panel on Climate Change to highlight the role of HEIs at global DRR efforts (Holloway, 2014). This indicates the role of HEIs in DRR is through their *expertise in DRR.*

Moreover, the *synergy between outcome of HEIs and its implementation* is vital for effective DRR strategies (Shaw et al., 2015). For example, during Cyclone Nargis, the agencies received EW information 3 days previously but the EW information was not delivered to people at risk. This shows that the capacity to identify risk indicators was there. However, the knowledge has not been transferred to relevant agencies for its implementation. Similarly, the case of the 2004 Indian Ocean Tsunmai incident, more than two and half hours were there to evacuate people. However, due to inability to communicate to people at risk, the largest number of human loss was recorded. This clearly highlights the availability of knowledge to identify risk was there. However, its implementation was lacking or delayed. This calls for urgent action to integrate educational products for implementation toward effective DRR through establishing synergy between research, education and implementation.

4.3 Role of higher-education institutions in coastal resilience in Asia

The role of HEIs for disaster resilience education in Asia is vital due to increasing hazard vulnerability in Asia (Malalgoda et al., 2014). Most disaster management related courses were introduced in universities after 2004 Indian Ocean Tsunami incident (Shaw et al., 2012).

According to Mujiburrahman (2018), the role of higher education institutions in Asia toward strengthening EW systems are in six forms. Their role is to assist in *advocacy* to the government in achieving the SFDRR. The second role of HEIs is to *supply necessary human resources* to technical agencies to provide risk information and monitoring. The third role is to *support the government in conducting risk assessment* to identify disaster prone areas in the country. The fourth role of HEIs in EW systems in Indonesia is *to train disaster risk management institutions* in Indonesia specifically for EW, specifically, to train people to enhance capacities among officials at local governments. *Establishment of academic network* is identified as the fifth role of the HEIs in Indonesia. This is to help the EW system through scientific contributions. The sixth role has been identified as *collaboration between stakeholders engaging* in EW. This has been done through the signing of a number of MOUs between the National Agency for Disaster Countermeasure in Indonesia (BNPB) and universities in Indonesia.

Krishnamurthy and Kamala (2015) explain the role of HEIs in coastal resilience in Asia based on a study conducted in Tamil Nadu. A university near to the coastal belt in Tamil Nadu has introduced an elective course on Disaster Management in their curriculum realizing the importance of *awareness and preparedness measures* after the 2004 Indian Ocean Tsunami. The area is also subjected to frequent threats of flooding and cyclones. Later, Japan and Australia started some joint programs with the university on disaster management. The university's direct and significant contribution *toward trainings and capacity building* has enhanced coastal resilience in the coastline of Tamil Nadu.

In addition, the Asian University Network for Environment and Disaster Management (AUEDM) has been established in Asia to *share knowledge* related to environmental and disaster risk reduction with prominent Asian universities in 2008. This aimed to bridge the gap between *academic research and education with practice fields* (Holloway, 2014).

After considering all above roles related to HEIs in the development of MHEW, the study identified three distinctive roles of HEIs in coastal resilience:

(1) Education and awareness raising
(2) Advocacy support for evidence-based policy making
(3) Creation of knowledge through research

Hence, Section 4.4 presents a proposed framework to link HEIs in the development of MHEW in Asia for enhancing coastal resilience based on literature.

4.4 Linking HEIs in MHEW development

According to Fig. 4.2, three major roles of HEIs can be linked to three types of enablers of effective MHEW.

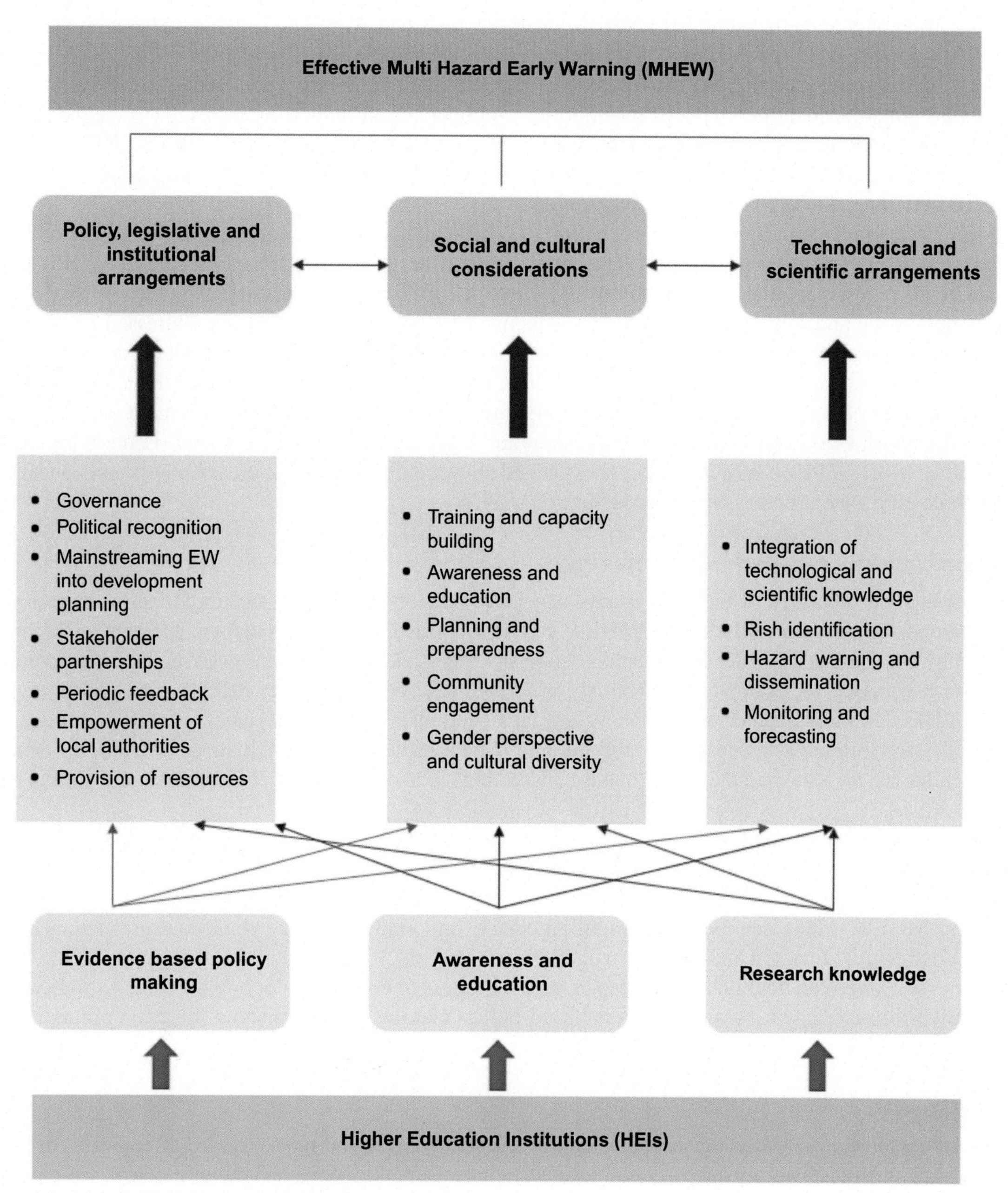

FIG. 4.2 Contribution of HEIs in developing MHEW.

Awareness and education

One of the main roles of HEIs is to provide education and raise awareness among stakeholders (Cortese, 2003). The government, policy makers, technical agencies, private sector, academics, media, NGOs and general public engage as stakeholders (Jubach and Tokar, 2016; Renn, 2012). Distribution of knowledge among many stakeholders to emphasize importance of MHEW, ways of identifying risks, and how to communicate risks. In addition, these awareness programs can be conducted among agencies, schools and communities which can create preparedness culture in the society (Thomalla and Larsen, 2010; UNISDR, 2015).

In addition, HEIs can contribute in many ways as described earlier to conduct training and capacity building among officials and the community for the development of MHEW (Mujiburrahman, 2018). Furthermore, through awareness and education and training, HEI can contribute to empower communities, and to promote active engagement in the design and implementation of MHEW (Collins, 2009b). These aspects are identified as the social and cultural consideration of effective MHEW. This shows that HEIs roles in education and awareness can be linked clearly for enhancing MHEW.

In addition, awareness and education can assist the enablers identified within policy, legislative and institutional arrangements. Through enhanced education and awareness, governance can be ensured when developing MHEW. Further, political support can be increased through awareness and education among political authorities as well as communities. Strong community support can influence political authorities to cooperate and support strategies that serve the communities most. According to the SFDRR, it is important to mainstream EW into development planning. Hence, HEIs can contribute through awareness and education programs for policy makers and practitioners introducing the importance and the ways in which this mainstreaming can be done (Zia and Wagner, 2015).

Shaw et al. (2012) pointed out that HEIs' contribution to DRR could be through educational knowledge creation, conducting research and contributing to practice of DRR. For example, education and awareness of coastal hazards and coastal resilience can be given to policy makers, community as well as public. Through education, awareness raising and knowledge creation most social and cultural consideration for effective MHEW can be improved or addressed.

Most local governments face many challenges while contributing to DRR. For example, lack of hazard and risk maps at local level, human resource limitations, insufficient funding, and political support are some of the major issues reported from local authorities in Sri Lanka (Malalgoda and Amaratunga, 2015). As a result, empowerment and capacity development of local authorities through training and education are suggested by many researchers (Lauterjung et al., 2010; UNESCO-ICO; UN/ISDR/PPEW; WMO, 2005). As explained above, the roles of HEIs on awareness and education hence contribute to three pillars identified as enablers of effective MHEW.

Advocacy and evidence-based policy making

The second role of HEIs identified by the study is advocacy and evidence-based policy making. In terms of policy, legislative and institutional arrangements for effective MHEW can be supported by HEIs through engagement and coordination. As highlighted by Basher (2006), better coordination between science and policy enhances governance. In addition, other related issues in governance for example, power struggle, lack of evidence-based policy making can be addressed through participation of HEIs. Another enabler within policy, legislative and institutional arrangement is lack of political support. This can be addressed through awareness programs among political leaders with the initiatives of HEIs. Evidence shows that HEIs can influence the state and the policy making institutions to avoid ad-hoc decision making by politicians (Shaw et al., 2012).

Through their advisory services, effective planning and preparedness strategies can be developed. Muttarak and Pothisiri (2013) explained the role of higher education for enhancing the level of preparedness among communities. Furthermore, experts in HEIs has the ability to highlight the importance of gender perspective and cultural diversity when preparing MHEW. Many scholars have highlighted the importance of gender and cultural perspective when designing DRR strategies (Enarson and Chakrabarti, 2009; Islam et al., 2017; Rogers and Tsirkunov, 2011). Hence experts' advice in such areas can help planning preparedness measures to consider in policy making. As indicated above, HEIs contribution through advocacy and evidence-based policy making for establishing all three pillars of effective MHEW is evident.

Conducting research

HEIs can contribute to achieve technological and scientific research support and inputs toward effective MHEW. In addition to their advice and consultancies in policy and legislative arrangements, social and cultural consideration are further highlighted.

Research conducted by HEIs provides input in many ways. One important aspect is to provide feedback evaluation. Research conducted by HEIs may evaluate efficiency and effectiveness of previous DRR strategies. These inputs can be used to recommend necessary changes through policies and legislative frameworks to enhance existing MHEW (Jubach and Tokar, 2016). In addition, engagement of HEIs promotes scientific evidence for policy making. HEIs can contribute in many other ways through their research. For example, risk assessment (Lauterjung et al., 2010), risk information dissemination, developing hazard warning and improving dissemination methods (Alfieri et al., 2012).

However, role of HEIs in DRR education has many challenges as highlighted by many researchers. HEIs generally provide programs that are delivered for achieving a specific qualification (Thayaparan et al., 2015). As a result, higher education is related with highly specialized areas of knowledge and traditional disciplines. Hence, traditional HEI's programs are fragmented into specialized disciplines without interdisciplinary interactions (Cortese, 2003) and with lack of collaboration or interactions with the industry (Thayaparan et al., 2015).

Hence, traditional formal education should go beyond university boundaries and engage with communities to learn from their experiences (Shaw et al., 2015). Thayaparan et al. (2015) argue that HEIs should be more innovative in providing opportunity to engage with many stakeholders such as industry, policy makers, communities, NGOs, private sector and other HEIs (Thayaparan et al., 2015). According to many researchers, however, this practice is still lacking in most of countries and regions (Shaw et al., 2011, 2015).

In addition to HEIs, strong regional cooperation is also recommended for enhancing resilience. Hence the next section presents the role of regional cooperation and its present status in Asia, and describes how the HEIs can assist for strong regional cooperation toward coastal resilience in Asia.

5 Importance of regional cooperation

According to Hollis (2015), more than 30 regional initiatives have been established across the world on disaster management during the last quarter of the 20th century. These initiatives have helped to enhance resilience capacities among member states. For example, the Association of Southeast Asian Nations (ASEAN) was set to help its neighbors to address economic, social, political and security issues. ASEAN deployed its Emergency Rapid Assessment Team (ASEAN-ERAT) to affected cities from the 2014 cyclone in the Philippines and coordinated with local authorities to provide relief support (Hollis, 2015).

Similar initiatives are reported in other regions of the world. The European Commission's Emergency Response Coordination Centre (ERRC) is another example where regional cooperation operates in the European region. It works under the EU Civil Protection Mechanism to coordinate and deliver assistance (relief assistance, expertise, relief protection teams and equipment) to countries affected by disasters in the European Region. This initiative also engages in prevention and preparedness measures for future hazards in the region (Europa, 2019). The African Union and the New Partnership for Africa's Development (NEPAD) work toward disaster risk reduction in Africa as a regional initiative.

Regional cooperation has been considered as vital when dealing with transboundary hazards (Seng, 2013). Most coastal hazards are transboundary in nature. For example, the 2004 Indian Ocean Tsunami affected 14 countries in the Asia and the Pacific Region. Similarly, Cyclone Haiyan affected mainly the Philippines and further, Southern China, Taiwan, etc. Tropical Cyclone Komen is another transboundary hazard which happened within this decade, in 2015, and which affected India, Nepal and Bangladesh (UNISDR, 2015).

Regional cooperation provides the opportunity to share scientific knowledge and applications, as well as to share costs when dealing with trans-boundary hazards (Haigh et al., 2018; Seng, 2013; UN-ESCAP, 2015). For example, Glacial Lake Outburst Floods (GLOF) initiative helps for risk assessment and mitigation along with sharing data and information among members to tackle transboundary glacial hazards in South Asia (Ives et al., 2010). Another example is the Indian Ocean Tsunami Warning System (IOTWS) which monitors and provides and information about hazard risks in Asia to enhance coastal resilience. Australia, India and Indonesia perform as Regional Tsunami Service Providers for the Indian Ocean providing timely warnings to countries in the region.

Many policy drivers require regional cooperation as an effective strategy for disaster risk reduction at global level. According to HFA framework, disaster risks must be integrated into development policies, plans and programs and supported through regional and international cooperation. It recognizes the development of EW as a responsibility of regional organizations which are responsible for implementing DRR. Furthermore the framework highlights the promotion of regional cooperation through transfer of knowledge, technology for capacity building, sharing of research findings, best practices and lessons learned (UNISDR, 2005). Later, the SFDRR, Section 34 [f] promoted the necessity of supporting regional cooperation toward better disaster preparedness. Section 34 [c] emphasizes the promotion and investment in effective nationally compatible regional MHEW mechanisms in line with global frameworks. In addition, sharing and exchange of information across countries is emphasized (UNISDR, 2015).

5.1 Existing regional cooperation/initiatives operate in Asia for coastal resilience

Having understood the benefit and applicability of regional cooperation for disaster risk reduction, especially for coastal risk reduction and resilience, some initiatives have been established after the Indian Ocean Tsunami incident in Asia. This is to improve people's access to timely and relevant warning information and to enable them to take life-saving actions as early as possible (Hettiarachchi, 2018; Thomalla and Larsen, 2010). Following are some of the initiatives established in the region.

(1) Establishment of the Indian Ocean Tsunami Warning & Mitigation System (IOTWMS) to provide rapid, accurate and reliable forecasting of future hazards in the Indian Ocean. They collaborate with Tsunami Service Providers (TSP) in India, Indonesia and Australia to monitor the Indian Ocean basin using state-of-the-art technology.

(2) The Indian Ocean Consortium was established to support the development of national components of the Indian Ocean Tsunami Warning Mitigation System (IOTWMS), through coordination mechanisms among governments, preparing national plans for tsunami warnings and creating links between regional efforts.

(3) With the development of IOTWS networks: Setting up deep ocean tsunameters sharing data in near real-time and available for warning purposes was increased to 100 in 2014 from 4 in 2004, broadband seismometers sharing data in near real-time from 13 to 140; coastline sea level gauges sharing data in real-time and available for tsunami warning purposes from 0 to 9

(4) The ASEAN Agreement on Disaster Management and Emergency Response (AADMER) was introduced in 2009 as the first legally binding instrument based on the Hyogo Framework for Action (HFA). This initiative was intended to promote cooperation and collaboration in reducing disaster losses and in undertaking joint emergency response. Its work program for 2010–2015 covered regional risk assessments, effective and efficient regional early warning activities, hazard monitoring to support mitigation efforts, and undertaking response and recovery activities. The succeeding work program for 2016–2020 still considers enhancing risk assessment and risk awareness as priority programs, but its emphasis is largely on protecting gains from community integration trough social protection programs and improving the resilience of infrastructures and essential services. The objective of this Agreement is to provide effective mechanisms to achieve substantial reduction of disaster losses in lives and in the social, economic and environmental assets of the Parties, and to jointly respond to disaster emergencies through concerted national efforts and intensified regional and international co-operation. This should be pursued in the overall context of sustainable development and in accordance with the provisions of this Agreement.

(5) The Indian Ocean Tsunami Information Centre was established in 2011 to support the common public in preventing, preparing and mitigating measures for tsunami across the Indian Ocean region. They provide information and resources, capacity building and conduct studies. It is operated through a web-based system to provide information, materials, data and programs.

(6) Setting up of Regional Integrated Multi-Hazard Early Warning Systems (RIMES) for Africa and Asia as an intergovernmental institution focusing the generation and application of EW information and building capacities in 2009. It aims to provide EW for tsunami and hydro-meteorological hazards. The consortium consists of 12 member states within Asia and Africa. Further, it provides facilities to enhance capacities of members to disseminate EW to communities at risk.

(7) The Severe Weather Forecasting Demonstration Project (SWFDP) operated in South-East Asia and South Asia. This project helps to strengthen capacity through conducting training to national meteorological and hydrological services in developing countries. This benefits the countries in advancement of science of weather forecasting, and also improves lead time and reliability for alerts and warning about future hazards

(8) The Sea Level Station Monitoring Facility receives data for monitoring sea level across 161 data providers established across the world. The Asia Disaster Preparedness Centre in Thailand, the China Meteorological Administration with three stations, the Department of Meteorology and Hydrology in Myanmar with three stations, the Department of Surveying and Mapping in Malaysia, the Geospatial Agency of Indonesia with 15 stations, the Gwadar Port Authority and Hydrographic Department in Pakistan, the Hong Kong Observatory in Hong Kong, the Hydrographic Department of Bangladesh Navy in Bangladesh are some of the monitoring stations in Asia.

(9) Regional specialized metrological centers were established in New Delhi and Tokyo. They operate within the framework of the WMO's World Weather Watch Programme and support countries with analysis, forecasting and training those who have limited capacities.

(10) Establishment of ASEAN Disaster Management and Emergency Relief Fund from voluntary contributions of its member states.

(11) Establishment of the Asian Disaster Preparedness Centre works as an intergovernmental organization and helps to build resilience to disasters in Asia and the Pacific region. It provides comprehensive technical services to countries in the region across social and physical sciences to support sustainable solutions for risk reduction and climate resilience. They further help in building regional and transboundary cooperation especially through capacity building and training programs.

However, regional cooperation toward coastal disaster risk reduction and resilience in Asia continue to suffer many challenges. Hettiarachchi (2018) pointed out that IOTWMS provides rapid accurate and reliable risk information for communities at risk. However, such systems need further technical advancement and support in assessing risks, developing warning centers and raising community awareness preparedness and resilience. For example, detailed inundation modeling and tsunami wave height forecasting are needed. He further highlighted the need to conduct studies on risk assessment on cities and community risk.

In this effort, training of officers to conduct such risk assessment are identified as an urgent requirement. Such activities are related to capacity building for countries, institutions as well as regional initiatives. Hettiarachchi (2018) further emphasizes the need to raise awareness among information suppliers and information receivers in the end-to-end EW system.

According to Evangelistaa et al. (2019) SWFDP provide accurate and reliable forecasts of temperature and thunderstorms for South East Asia. However, it is unable to forecast wind speed. This suggests the importance of capacity development among officers of institutions of regional initiatives toward effective MHEW.

To address the above identified gaps in the regional cooperation, HEIs can be integrated to strengthen regional cooperation for effective MHEW in Asia. This has been emphasized by Holloway (2014) showing how such HEIs enhance skills and capacities among officers who are involved in DRR. The Pacific region has shown a strong commitment to regional cooperation, cooperative decision making with strong political support along with technical support. This has provided remarkable improvement in disaster risk management and strong relationship between stakeholders in the Pacific region. They also have advanced institutional facilities which provide cost-effective academic research, technical education and information management. However, there is no such joint HE academic research and training specific to DR management in Asia compared to Africa and Latin America. For example, India offers six-seven universities offer master-level programs on disaster management while Columbia, Peru and South Africa offer three-four master programs on disaster risk management by each universities (Holloway, 2014).

6 Conclusions

Coastal hazards have become an emergent threat in Asia for lives and economic development. Many initiatives have been created to enhance coastal resilience. MHEW has been identified as an effective strategy to enhance coastal resilience in Asia. However, the present status of MHEW in Asia needs more improvements within all three pillars. In this effort, HEIs are identified as agents who can contribute through raising education and awareness, provisioning advocacy for evidence-based policy making and conducting research. In addition, the role of HEIs can also be found in Asia for strengthening regional cooperation. This is because, regional cooperation has the capacity to share knowledge and expertise as well as share cost of initiating and operating MHEW. Hence, the study has introduced an innovative way in which HEIs can contribute to the development of MHEW through strengthening regional cooperation based on literature. The study findings provide insights for further study and inputs for strengthening coastal resilience in Asia.

Acknowledgments

This paper is based on the initial stage of project CABARET which was supported by the European Commission Erasmus+CBHE Project 573816, "CApacity Building in Asia for Resilience EducaTion" (CABARET). The European Commission support for the production of this publication does not constitute an endorsement of the contents which reflects the views only of the authors, and the Commission cannot be held responsible for any use which may be made of the information contained therein.

References

Aitsi-Selmi, A., Murray, V., Wannous, C., Dickinson, C., Johnston, D., Kawasaki, A., Yeung, T., 2016. Reflections on a science and technology agenda for 21st century disaster risk reduction. Int. J. Disaster Risk Sci. 7 (1), 1–29.

Alam, K., Rahman, M.H., 2014. Women in natural disasters: a case study from southern coastal region of Bangladesh. Int. J. Disaster Risk Reduct. 8, 68–82.

Alfieri, L., Salamon, P., Pappenberger, F., Wetterhall, F., Thielen, J., 2012. Operational early warning systems for water-related hazards in Europe. Environ. Sci. Pol. 21, 35–49.

Basher, R., 2006. Global early warning systems for natural hazards: systematic and people-centred. Philos. Trans. R. Soc. Lond. A 364 (1845), 2167–2182.

Besset, M., Anthony, E.J., Dussouillez, P., Goichot, M., 2017. The impact of cyclone Nargis on the Ayeyarwady (Irrawaddy) River delta shoreline and nearshore zone (Myanmar): towards degraded delta resilience? Comptes Rendus Géosci. 349 (6), 238–247.

Brunkard, J., Namulanda, G., Ratard, R., 2005. Hurricane katrina deaths, Louisiana. Disaster Med. Public Health Prep. 2 (4), 215–223.

Buan, S., Diamond, L., 2012. Multi-hazard early warning system of the United States National Weather Service. In: Institutional Partnerships in Multi-Hazard Early Warning Systems. Springer, Chippenham, pp. 115–157.

Chadenas, C., Creach, A., Mercier, D., 2014. The impact of storm Xynthia in 2010 on coastal flood prevention policy in France. J. Coast. Conserv. 18 (5), 529–538.

Chaimanee, N., 2006. The role of geoscience in Hazard mitigation. Asian Disaster Manag. News Newsl. 12, 6.

Challinor, A., Wheeler, T., Garforth, C., Craufurd, P., Kassam, A., 2007. Assessing the vulnerability of food crop systems in Africa to climate change. Clim. Chang. 83 (3), 381–399.

CNN, 2018a. Indonesia Tsunami: Death Toll Rises Sharply as Desperation Grows. https://edition.cnn.com/2018/10/02/asia/indonesia-palu-tsunami-earthquake-intl/index.html. (Accessed 10 July 2019).

CNN, 2018b. Indonesia Tsunami: Grim Search for Survivors Continues as Death Toll Reaches 430. CNN World. https://edition.cnn.com/2018/12/25/asia/indonesia-tsunami-intl/index.html. (Accessed 8 July 2019).

Collins, A.E., 2009a. Disaster and Development. Routledge, UK.

Collins, A.E., 2009b. Early Warning: A People-Centred Approach to Early Warning Systems and the 'Last Mile'. International Federation of Red Cross and Red Cresent Societies.

Cortese, A.D., 2003. The critical role of higher education in creating a sustainable future. Plan. High. Educ. 31 (3), 15–22.

CRED, 2018. Economic Losses, Poverty and Disasters 1998–2017. The Centre for Research on the Epidemiology of Disasters, p. 52.

CRED and UNISDR, 2016. Tsunami Disaster Risk: 2016 Past impacts and Projections. The Centre for Research on the Epidemiology of Disasters.

CRED-UNISDR, 2016. Poverty & Death: Disaster Mortality. The Centre for Research on the Epidemiology of Disasters and United Nations Office for Disaster Risk Reduction.

Davis, J.L., Currin, C.A., O'Brien, C., Raffenburg, C., Davis, A., 2015. Living shorelines: coastal resilience with a blue carbon benefit. PLoS ONE 10 (11), e0142595.

Dutta, R., Basnayake, S., 2018. Gap assessment towards strengthening early warning systems. Int. J. Disaster Resil. Built. Environ. 9 (2), 198–215.

Enarson, E.P., Chakrabarti, P.G.D., 2009. Women, Gender and Disaster: Global Issues and Initiatives. SAGE, Los Angeles, CA; London.

Euroeducation, 2014. England Higher Education System. https://www.euroeducation.net/prof/ukco.htm. (Accessed 20 July 2019).

Europa, 2019. EU Civil Protection Mechanism. https://ec.europa.eu/echo/what/civil-protection/mechanism_en. (Accessed 15 June 2019).

Evangelistaa, C.J.V., Esquivelb, V.S., Nimesb, J.M., Guidoa, R.M.D., 2019. Evaluation of Severe Weather Forecasting Demonstration Project–Southeast Asia (SWFDP-Sea) Weather Forecast for Selected Weather Parameters.

Frankenberg, E., Sikoki, B., Sumantri, C., Suriastini, W., Thomas, D., 2013. Education, vulnerability, and resilience after a natural disaster. Ecol. Soc. 18 (2), 16.

Gall, M., Cutter, S.L., Nguyen, K., 2014. Governance in Disaster Risk Management. Retrieved from Beijing.

Golnaraghi, M., 2012. Institutional Partnerships in Multi-Hazard Early Warning Systems: A Compilation of Seven National Good Practices and Guiding Principles. Springer Science & Business Media, pp. 1–8.

Gornitz, V., 1991. Global coastal hazards from future sea level rise. Glob. Planet. Chang. 3 (4), 379–398.

Habib, A., Shahidullah, M., Ahmed, D., 2012. The Bangladesh cyclone preparedness program. A vital component of the nation's multi-hazard early warning system. In: Institutional Partnerships in Multi-Hazard Early Warning Systems. Springer, pp. 29–62.

Haigh, R., Amartunga, D., Hemachandra, K., 2018. A capacity analysis framework for multi-hazard early warning in coastal communities. Proc. Eng. 212, 1139–1146.

Harley, M.D., Valentini, A., Armaroli, C., Perini, L., Calabrese, L., Ciavola, P., 2016. Can an early-warning system help minimize the impacts of coastal storms? A case study of the 2012 Halloween storm, northern Italy. Nat. Hazards Earth Syst. Sci. 16 (1), 209–222.

Hasegawa, N., Harada, S., Tanaka, S., Ogawa, S., Goto, A., Sasagawa, Y., Washitake, N., 2012. Multi-Hazard early warning system in Japan. In: Institutional Partnerships in Multi-Hazard Early Warning Systems. Springer, pp. 181–215.

Hatakenaka, S., 2015. The role of higher education institutions in innovation and economic development. Int. Higher Educ., 47.

Healy, A., Malhotra, N., 2009. Myopic voters and natural disaster policy. Am. Polit. Sci. Rev. 103 (3), 387–406.

Hemachandra, K., Amaratunga, D., Haigh, R., 2018. Role of women in disaster risk governance. Proc. Eng. 212, 1187–1194.

Hemachandra, K., Haigh, R., Amaratunga, D., 2019. Regional cooperation towards effective multi-hazard early warnings in Asia. Int. J. Adv. Sci. Eng. Inf. Technol. 9 (1), 287–292.

Hettiarachchi, S., 2018. Establishing the Indian Ocean tsunami warning and mitigation system for human and environmental security. Proc. Eng. 212, 1339–1346.

Hollis, S., 2015. The role of regional organizations in disaster risk management. In: The Role of Regional Organizations in Disaster Risk Management. Springer, pp. 1–12.

Holloway, A., 2014. Srategic Mbilisation of Higher Education Institutions in Disaster Risk Reduction Capacity Building: Expereince of Periperi U. Retrieved from South Africa.

Hugo, G., 2011. Future demographic change and its interactions with migration and climate change. Glob. Environ. Chang. 21, S21–S33.

IFRC, 2009. World Disasters Report: Focus on Early Warning, Early Actions. Retrieved from Satigny/Vernier, Switzerland. The International Federation of Red Cross and Red Crescent Societies.

IGI Global, 2019. What is Higher Education Institutions (HEIs). Retrieved from: https://www.igi-global.com/dictionary/information-system-projects-for-higher-education-management/13099. (Accessed 14 August 2019).

IPCC, 2014. Summary for policymakers. In: Climate Change 2014: Impacts, Adaptation, and Vulnerability. Part A: Global and Sectoral Aspects. Contribution of Working Group II to the Fifth Assessment Report of the Intergovernmental Panel on Climate Change. Cambridge University Press, Cambridge, United Kingdom and New York, NY, USA. Retrieved from:.

Ishiwatari, M., 2012. Government roles in community based disaster risk reduction. In: Shaw, R. (Ed.), Community Based Disaster Risk Reduction. Emerald Group Publishing.

Islam, M.R., Ingham, V., Hicks, J., Manock, I., 2017. The changing role of women in resilience, recovery and economic development at the intersection of recurrent disaster: a case study from Sirajgang, Bangladesh. J. Asian Afr. Stud. 52 (1), 50–67.

Ives, J.D., Shrestha, R.B., Mool, P.K., 2010. Formation of Glacial Lakes in the Hindu Kush-Himalayas and GLOF risk Assessment. ICIMOD, Kathmandu, Nepal.

Jubach, R., Tokar, A., 2016. International severe weather and flash flood hazard early warning systems—leveraging coordination, cooperation, and partnerships through a hydrometeorological project in Southern Africa. Water 8 (6), 258.

Kadel, M., 2011. Community Participation in Disaster Preparedness Planning: A Comparative Study of Nepal and Japan. Asian Disaster Reduction Centre Final Report.

Kelman, I., Glantz, M.H., 2014. Early warning systems defined. In: Reducing Disaster: Early Warning Systems for Climate Change. Springer, pp. 89–108.

Koehn, P., 2013. Donor-supported transnational higher education initiatives for development and research: a framework for analysis and a call for increased transparency. Higher Educ. Policy 26 (3), 349–372.

Kogan, M., Becher, T., 1980. Process and Structure in Higher Education. Heinemann Educational Books, London.

Kottegoda, S., 2011. Mainstreaming Gender in Disaster Management Policy: Key Issues and Challenges in the Asia-Pacific Region. http://www.apww-slwngof.org/index.php?option=com_content&view=article&id=88. (Accessed 23 May 2019).

Krishnamurthy, R., Kamala, K., 2015. Impact of Higher Education in Enhancing the Resilience of Disaster Prone Coastal Communities: a case study in Nemmeli Panchayat, Tamil Nadu, India. In: Recovery from the Indian Ocean Tsunami. Springer, pp. 361–380.

Lauterjung, J., Münch, U., Rudloff, A., 2010. The challenge of installing a tsunami early warning system in the vicinity of the Sunda Arc, Indonesia. Nat. Hazards Earth Syst. Sci. 10 (4), 641.

Liyanage, C.L., Thakore, R., Amaratunga, D., Mustapha, A., Haigh, R., 2018. The barriers to research and innovation in disaster resilience in higher education institutions in Asia. Proc. Eng. 212, 1225–1232.

Malalgoda, C., Amaratunga, D., 2015. A disaster resilient built environment in urban cities: the need to empower local governments. Int. J. Disaster Resil. Built. Environ. 6 (1), 102–116.

Malalgoda, C., Amaratunga, D., Haigh, R., 2014. Challenges in creating a disaster resilient built environment. Proc. Econ. Finance 18, 736–744.

Mayhorn, C.B., McLaughlin, A.C., 2014. Warning the world of extreme events: a global perspective on risk communication for natural and technological disaster. Saf. Sci. 61, 43–50.

Meerpoël, M., 2015. Disaster Risk Governance: The Essential Linkage Between DRR and SDGs. Faculty of Law, Catholic University of Lille France.

Merkens, J.-L., Reimann, L., Hinkel, J., Vafeidis, A.T., 2016. Gridded population projections for the coastal zone under the shared socioeconomic pathways. Glob. Planet. Chang. 145, 57–66.

Mori, N., Takemi, T., 2016. Impact assessment of coastal hazards due to future changes of tropical cyclones in the North Pacific Ocean. Weather Clim. Extrem. 11, 53–69.

Mujiburrahman, M., 2018. The role of universities in Indonesia multi-hazard early warning system. MATEC Web Conf. 229, 1–8.

Muttarak, R., Pothisiri, W., 2013. The role of education on disaster preparedness: case study of 2012 Indian Ocean earthquakes on Thailand's Andaman coast. Ecol. Soc. 18 (4), 1–16.

Nakahara, S., Ichikawa, M., 2013. Mortality in the 2011 tsunami in Japan. J. Epidemiol., JE20120114.

NDRRMC, 2013. NDRRMC Update: Final Report re Effects of Typhoon 'Yolanda' (Haiyan). The National Disaster Risk Reduction and Management Council. Retrived from: http://ndrrmc.gov.ph/attachments/article/1329/final_report_re_effects_of_typhoon_yolanda_(haiyan)_06-09NOV2013.pdf. (Accessed 13 April 2019).

Paul, B.K., 2009. Why relatively fewer people died? The case of Bangladesh's cyclone Sidr. Nat. Hazards 50 (2), 289–304.

Perdikou, S., Horak, J., Palliyaguru, R., Halounová, L., Lees, A., Ranguelov, B., Lombardi, M., 2014. The current landscape of disaster resilience education in Europe. Proc. Econ. Finance 18, 568–575.

Pew Research Center for the People and the Press, 2013. Trust in Government Nears Record Low, But Most Federal Agencies Are Viewed Favorably. http://www.people-press.org/2013/10/18/trust-in-governmentnears- record-low-but-most-federal-agencies-are-viewed-favorably/. (Accessed 28 April 2019).

Renn, O., 2012. Inclusive Risk Governance. 2011–2012 Distinguish Lecture Series, Environmental Scienceand Policy Programme, Michigen State University.

Rogers, D., Tsirkunov, V., 2011. Implementing hazard early warning systems. In: Global Facility for Disaster Reduction and Recovery.

Saito, F., 2012. Women and the 2011 East Japan disaster. Gend. Dev. 20 (2), 265–279.

Scott, Z., Few, R., 2016. Strengthening capacities for disaster risk management: insights from existing research and practice. Int. J. Disaster Risk Reduct. 20, 145–153.

Seidel, H., 1991. Internationalisation: a new challenge for universities. High. Educ. 21 (3), 289–296.

Seng, D.S.C., 2013. Tsunami resilience: multi-level institutional arrangements, architectures and system of governance for disaster risk preparedness in Indonesia. Environ. Sci. Pol. 29, 57–70.

Seto, K.C., Fragkias, M., Güneralp, B., Reilly, M.K., 2011. A meta-analysis of global urban land expansion. PLoS ONE 6 (8).

Setyono, J.S., Yuniartanti, R.K., 2016. The challenges of disaster governance in an Indonesian multi-hazards city: a case of Semarang, Central Java. Procedia Soc. Behav. Sci. 227, 347–353.

Shaw, R., Mallick, F., Takeuchi, Y., 2011. Essentials of higher education in disaster risk reduction: prospects and challenges. In: Disaster Education. Emerald Group Publishing Limited, pp. 95–113 (Chapter 5).

Shaw, R., Takeuchi, Y., Krishnamurthy, R., Pereira, J.J., Mallick, F., 2012. Universities and community based disaster risk reduction. Commun. Environ Disaster Risk Manag. 10, 55–66.

Shaw, R., Ishiwatari, M., Arnold, M., 2015. Community-Based Disaster Risk Management.

Spalding, M.D., Susan Ruffo, C.L., Imen Meliane, L.Z.H., Shepard, C.C., Beck, M.W., 2014. The role of ecosystems in coastal protection: adapting to climate change and coastal hazards. Ocean Coast. Manag. 90, 50–57.

Strunz, G., Post, J., Zosseder, K., Wegscheider, S., Mück, M., Riedlinger, T., Gebert, N., 2011. Tsunami risk assessment in Indonesia. Nat. Hazards Earth Syst. Sci. 11 (1), 67–82.

Tang, X., Feng, L., Zou, Y., Mu, H., 2012. The Shanghai multi-hazard early warning system: addressing the challenge of disaster risk reduction in an urban megalopolis. In: Institutional Partnerships in Multi-Hazard Early Warning Systems. Springer, pp. 159–179.

Thayaparan, M., Siriwardena, M., Malalgoda, C.I., Amaratunga, D., Lill, I., Kaklauskas, A., 2015. Enhancing post-disaster reconstruction capacity through lifelong learning in higher education. Disaster Prev Manag 24 (3), 338–354.

Thomalla, F., Larsen, R.K., 2010. Resilience in the context of tsunami early warning systems and community disaster preparedness in the Indian Ocean region. Environ. Hazards 9 (3), 249–265.

UN, 2015. Sustainable Development Goals. https://sustainabledevelopment.un.org/?menu=1300. (Accessed 24 July 2019).

UN-ESCAP, 2015. Strengthening Regional Multi-Hazard Early Warning Systems. The United Nations Economic and Social Commission for Asia and the Pacific, Bangkok.

UNESCO-ICO; UN/ISDR/PPEW; WMO, 2005. Assessment of Capacity Building Requirements for an Effective and Durable Tsunami Warning and Mitigation System in the Indian Ocean: Consolidated Report for 16 Countries Affected by the 26 December 2004 Tsunami. The Intergovernmental Oceanographic Commission of UNESCO, UN/ISDR Platform for the Promotion of Early Warning and World Meteorological Organization, Paris.

UNFCCC, 2015. Paris Agreement. The United Nations Framework Convention on Climate Change. https://unfccc.int/process-and-meetings/the-paris-agreement/the-paris-agreement. (Accessed 20 May 2019).

UNISDR, 2003. Conference Statement: Second International Conference Early Warning II. United Nations Office for Disaster Risk Reduction, Bonn, Germany.

UNISDR, 2005. Hyogo Framework for Action 2005–2015: Building the Resilience of Nations and Communities to Disasters. United Nations Office for Disaster Risk Reduction, Geneva, Switzerland.

UNISDR, 2006. Final Statement Third International Conference on Early Warning (EWC III). United Nations Office for Disaster Risk Reduction, Bonn, Germany. Retrieved from, https://www.unisdr.org/2006/ppew/info-resources/ewc3/FinalStatementfinal.pdf.

UNISDR, 2009. UNISDR Terminology. United Nations Office for Disaster Risk Reduction, Geneva, Switzerland. https://www.undrr.org/publication/2009-unisdr-terminology-disaster-risk-reduction.

UNISDR, 2015. Sendai Framework for Disaster Risk Reduction 2015–2030. United Nations Office for Disaster Risk Reduction, Geneva, Switzerland. http://www.preventionweb.net/files/43291_sendaiframeworkfordrren.pdf.

UNISDR, 2017. UNISDR Terminology on Disaster Risk Reduction. United Nations Office for Disaster Risk Reduction, Geneva, Switzerland.

Valenzuela, M., Gangai, J., 2008. Coastal Geofirm tools for coastal risk identification. In: Solutions to Coastal Disasters, pp. 410–421.

WMO, 2016. Multi-Hazard Early Warning Systems (MHEWS). World Metereological Organization. http://www.wmo.int/pages/prog/drr/projects/Thematic/MHEWS/MHEWS_en.html. (Accessed 14 May 2019).

WMO, 2017. Good Practices for Multi-Hazard Early Warning Systems (EWS). World Metereological Organization. https://public.wmo.int/en/resources/meteoworld/good-practices-multi-hazard-early-warning-systems-ews. (Accessed 17 February 2017).

WMO, 2018. Multi-hazard Early Warning Systems: A Checklist. World Metereological Organization, Geneva, Switzerland.

World Economic Forum, 2017. Realizing Human Potential in the Fourth Industrial Revolution An Agenda for Leaders to Shape the Future of Education, Gender and Work. World Economic Forum, Switzerland.

Zia, A., Wagner, C.H., 2015. Mainstreaming early warning systems in development and planning processes: multilevel implementation of Sendai framework in Indus and Sahel. Int. J. Disaster Risk Sci. 6 (2), 189–199.

Zou, L.-L., Wei, Y.-M., 2010. Driving factors for social vulnerability to coastal hazards in Southeast Asia: results from the meta-analysis. Nat. Hazards 54 (3), 901–929.

Chapter 5

Using financial instruments and PPP schemes for building resilience to natural disasters

Felix Villalba-Romero[a] and Champika Liyanage[b]
[a]*EAE Business School, Barcelona, Spain,* [b]*University of Central Lancashire, Lancashire, United Kingdom*

1 Introduction

Traditionally, consequences and damages occur from natural disasters/hazards have been managed after the event takes place and political leaders have obtained consensus to set priority for the use of resources to deal with the disasters. This also includes engaging people and institutions through a plethora of collaborative forms. Currently, due to climate change issues, natural disasters are mostly recurrent and, therefore, require preparation of the society to have a greater capacity of resistance to minimize the damages caused by the disasters through Disaster Risk Reduction (DRR) and Disaster Risk Management (DRM) tools. The former refers to the concept and practice of reducing disaster risks through systematic efforts in order to analyze and manage the causal factors of disasters. DRR is a response to the challenges of climate change and, in particular, it needs to be addressed by an efficient disaster risk management (DRM) process, which refers to a "systematic process of using administrative directives, organizations, and operational skills and capacities to implement strategies, policies and improved coping capacities in order to lessen the adverse impacts of hazards and the possibility of disaster" (UNISDR, 2009). The economic impact of the climate change induced disasters/hazards has led to the introduction of Disaster Risk Financing (DRF) to cover the direct and indirect economic losses incurred by the society. DRF is under the umbrella of DRM and it covers the financial management of DRM actions with the use of necessary Financial Instruments (FI) (Clarke et al., 2015a,b). Thus, FI is a key part of DRF strategy. In this context, it is necessary to identify characteristics that enhance conventional FI or that give rise to innovative FI that are applicable for DRR.

It is apparent from the above that terms DRM, DRR, DRF and FI are very much interlinked and, thus needs careful consideration of what each term means and what each of their roles are in the broader disaster management concept. The purpose of this research is to partly fulfill this purpose by identifying the role of DRF and FI with the use of a systematic review process.

2 Methodology

A systematic review process was performed to identify DRF related knowledge and experiences that may contextualize the current state of art and guide the area of potential development in terms of financial instruments. This review can be defined as a process of "synthesizing research in a systematic, transparent and a reproducible manner" (Tranfield et al., 2003). The researchers can carry out the following steps: (1) identification of the field of study and the period to be analyzed; (2) selection of the information sources; (3) search; (4) management and debugging of search results; and (5) analysis of results (Alfalla-Luque et al., 2013).

Initially a google scholar advance search on disaster risk financing was carried out, only in the title of the article. The reason was to specifically find articles that have discussed/researched on the area of DRF. The same search was carried out in other popular social science databases such as Emerald, Science Direct, Taylor and Francis and Web of Science; in order to ensure wide coverage of the search term. Altogether, the research yielded around 114 documents, i.e., 65 on Google Scholar and 39 on other databases (see Fig. 5.1). Most of the 114 documents were published within the last 3–5 years, which may indicate the increased attention given to the area of research in recent years. Most of the publications were from relevant academic databases and resources of public institutions aiming to promote DRR (mainly UN, The World Bank Group,

Strengthening Disaster Risk Governance to Manage Disaster Risk. https://doi.org/10.1016/B978-0-12-818750-0.00005-2

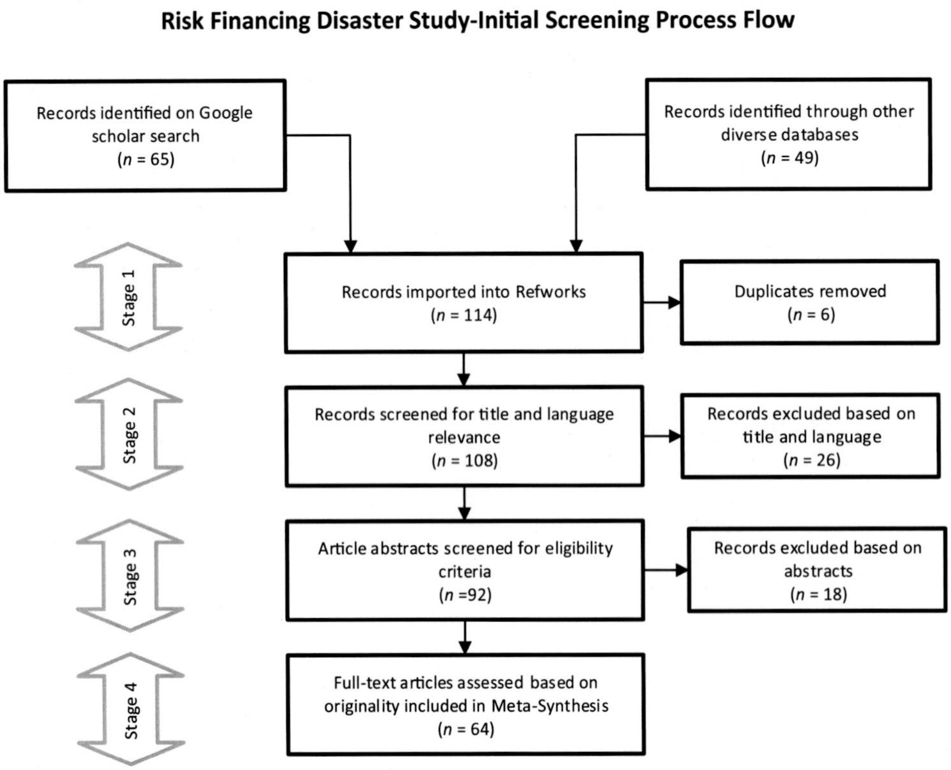

Risk Financing Disaster Study-Initial Screening Process Flow

Records identified on Google scholar search (*n* = 65)

Records identified through other diverse databases (*n* = 49)

Stage 1

Records imported into Refworks (*n* = 114)

Duplicates removed (*n* = 6)

Stage 2

Records screened for title and language relevance (*n* = 108)

Records excluded based on title and language (*n* = 26)

Stage 3

Article abstracts screened for eligibility criteria (*n* =92)

Records excluded based on abstracts (*n* = 18)

Stage 4

Full-text articles assessed based on originality included in Meta-Synthesis (*n* = 64)

FIG. 5.1 Systematic review screening process.

and other multilateral developing institutions leading the area of Disaster Risk Reduction). This assures the authenticity of the documents selected for the systematic review.

Of the 114 documents, 6 were initially removed due to duplications and further 44 documents were removed after considering language of the paper (only documents written in English were chosen) and screening the title and abstract of the paper. Since the analysis was carried out qualitatively, the "relevance" of the literature to the subject matter was very important. The selected 64 documents were then stored in excel and refworks. The softwares were also used to record the publication data and to record the analysis of the documents in-depth. The analysis of the documents focused on the main research questions of the systematic review; (1) What is DRF?; (2) What Financial Instruments, both traditional and innovative, can be used as part of DRF strategies?; (3) Who are the actors that can implement DRF strategies?; (4) What stages does DRF need to be considered?; (5) How should DRF be implemented and in what ways can it be applied? This paper only presents answers to the first two research questions above and they are presented in the following sections.

3 Disaster risk financing

The financial impact of disasters can be mitigated using appropriate financial management and risk transfer tools. These tools may be planned and implemented ex-ante and may include compensation arrangements provided by the private sector or the government insurance coverage. All these tools that provide financial protection may be included in the global concept of Disaster Risk Financing (DRF). Although there is not a specific definition of the term, DRF consists of financial strategies and instruments that manage the financial impact of disasters, mitigating the cost of disaster risks and reducing the financial burden, as well as contributing to a rapid economic recovery (OECD/G20, 2012). Herein, a financial strategy is integral to an organization's plan that summarizes targets, actions and funding requirements to be taken over the period considered for disaster planning. Developing a financial strategy for governments includes a number of strategic options for financing rather than just funding from public budget and donors, in order make the strategy/ies more sustainable. Moreover, innovative solutions for raising money to cover specific risks are required and generally could be structured through ad hoc financial instruments which are recently being developed.

DRF, a derivation of risk financing, determines how an organization (e.g., government or the public sector) will pay for losses incurred from disasters in the most effective and least costly way possible. For this, there is a need to carryout effective risk assessment of pre-disaster, during disaster, post-disaster recovery stages. Risk financing is concerned with providing funds to cover the effect of unexpected losses, and the main finance forms include risk transfer, risk retention funded by reserves (self-insurance) and risk pooling which provides protection to the vulnerable communities. Risk financing, moreover, involves the identification of risks, the financial allocation of each of the risks and monitoring the effectiveness of the selected financial instrument to cover each risk. The management of the risks, of which risk financing is an integral part, is the set of measurable and sustainable actions for reducing the effect of the uncertainty.

OECD defines a methodological framework for risk assessment and risk financing at different stages of a disaster (OECD/G20, 2012). In a pre-disaster phase, elements such as governance, hazards, and the level of exposure and vulnerability determines the risk level, and in a post disaster phase, the impact should be analyzed. This assessment provides information for prevention, mitigation and emergency preparedness when defining disaster risk financing schemes, there is a need to evaluate financial exposure, the type and level of risks to be transferred and risk financing strategy and institutional arrangements. Following the whole process will enable policy makers to allocate the optimal financial instrument for the specific identified risks.

The risks are generally assessed by damages multiplied by the frequency of the hazard. Not all damages may be easily quantified specially in term of victims and social costs. However, economical losses are usually categorized under households, business and public infrastructures. In addition, agricultural losses derived from hazards and adverse weather conditions are generally considered as well on top of the above. Main hazards which cause major losses could be in the form of earthquakes, tsunamis, floods, volcanic eruptions, atypical cyclonic storm and meteorite falls. The risks of each hazard require a different and appropriate DRF strategy and the use of a different financial instrument. Therefore, DRF aims to achieve financial resilience by setting the financial strategy and selecting the most appropriate financial instruments. One of the most effective way is achieved when the Government and the private sector merges using integrated DRF tools (OECD, 2015).

3.1 Government actions on DRF

In view of Government, they should rely on (1) compensations, (2) financial assistance arrangements, both aiming to address financial vulnerability where the aforementioned private coverage is insufficient or not affordable to some communities, and (3) sovereign risk financing strategies. These strategies may be ex-ante approaches, such as mobilizing reserve funds, making contingent credit arrangements and implementing risk transfer tools; or ex-post approaches, such as borrowing from financial markets, funding allocations and taxation.

Government compensation and financial assistance arrangements, as ex-ante measures, should be clearly defined. The specific parameters on which those measures are based, should also be included, mostly in terms of prompt assistance. This is essential to ensure financial protection to the most vulnerable parts of the society, but also to avoid double payments or disincentives for self-protection. This way, it avoids crowding out private initiative in risk reduction.

Regional scope and cooperation for financial assistance is another interesting ex-ante approach to be explored, e.g., EU Solidarity Fund. This fund was established in 2002 to assist EU member states in recovering from natural disasters, mostly providing prompt assistance for evacuation and temporary re-accommodation, emergency services and clean-up for disaster victims or areas, as well as, covering not insurable losses. Other examples of regional initiatives are the risk pooling through the Pacific Disaster Risk Financing and Insurance (PDRFI), the Caribbean Catastrophe Risk Insurance Facility (CCRIF) and the African Risk Capacity (FRC) (OECD, 2015).

Sovereign Disaster Risk Financing (SDRF) is another approach to DRF. It aims to increase the capacity of governments in providing immediate emergency funding as well as long-term funding for reconstruction and development (World Bank Group, 2014). Sovereign Disaster Risk Financing and Insurance (SDRFI) programs also offer public financial management and implementation of market-based sovereign catastrophe risk insurance solutions especially for post-disaster phase (Ley-Borrás and Fox, 2015; Ghesquiere and Mahul, 2007; De Janvry, 2015; Teh, 2014, 2015; Boudreau, 2015).

In order to mitigate financial impacts of a disaster, governments may complement the investments in risk reduction with ex ante DRF tools. These ex-ante financial tools may address short-term (emergency response), mid-term (recovery) or long-term (reconstruction) disaster impacts, which can be used in combination. Most common ex ante or pre-disaster DRF tools are:

- Government reserves such as dedicated contingency funds for disasters or multi-year disaster reserve funds.
- Contingent credit arrangements with a financial institution or international organization.
- Actions suitable for risk transfer such as, catastrophe bonds or other types of catastrophe-linked securities or derivatives, as well as, insurance promotion to cover indemnities against damage.

The main challenges to implement Government actions in DRF are diverse. In terms of compensation and financial assistance arrangements, it is difficult to allocate limited available financial resources among all disaster victims. This is mainly due to lack of clarity in terms of the allocation responsibility between the public and private actors; and difficulties in ensuring transparency and accountability, while setting a speedy compensation. As for the challenges in implementing sovereign risk disaster strategies, one of the main challenges is insufficient ex-ante commitment of financial resources. In addition, other challenges are lack of information, documentation and coordination, due to the absence of a proper legal and regulatory framework that enable the implementation of market-based sovereign risk transfer mechanisms.

3.2 Private sector actions on DRF

Herein, the main financial instrument is the private insurance, which play a key role in mitigating the costs of hazard disasters. In most developed countries, private insurance coverage is broad and become the main financial tool for business and households to absorb the financial impact of hazards. Related insurance and reinsurance markets are extensive and allow a big portion of risk to be transferred to the private sector. However, in some economies, disaster coverage is limited due to the reduced scope of the insurance markets, while in other countries, the availability and affordability of disaster insurance is the main issue. Another important matter is the public support, in terms of guarantees or reinsurance schemes.

The challenges in the implementation of the private insurance are mainly the constrains in terms of both the demand for insurance and the ability to provide insurance option economically viable. But there are also other important ones, such as lack of insurance culture associated with uncertainty of the claim payments, product complexity in the case of innovative index-based or parametric insurance products, specific challenges for the agricultural insurance due to the frequency of events that causes damages or losses, and especially the potential moral hazard because individuals and business expect to receive compensation from the government.

The OECD in its G20 leaders' mandate drew up a report on Methodological Framework for Disaster Risk Assessment and Risk Financing, which recognizes the importance and priority of disaster risk management (DRM) strategies focusing on disaster risk assessment and risk financing (OECD/G20, 2012). The framework addresses risk assessment as the first key step for risk assessment strategies and include among the relevant tasks, the evaluation of the availability of risk financing and risk transfer tools to address the financial impact of disasters. This requires the identification of the scope for risk retention and transfer, and the assessment of cost and benefit of the available financial tools, particularly the insurance instruments that may provide post-disaster aid.

Private sector insurers have developed innovative financial risk transfer products to mitigate the impact of disaster events that contribute to reducing the burden on public budgets (Swiss Re, 2008). Usually, insurance mechanisms do not cover the full amount of the damage and there is a funding gap that eventually is supported by the people with some relief funds by the governments (Ozaki, 2016), which have been evaluated in some reports (Swiss Re, 2008) and require innovative solutions (Swiss Re, 2011). Modeling works to estimate losses derived from disaster also contribute to develop methodologies for quantifying the economic impact of disaster (Winspear et al., 2012; Clarke and Mahul, 2011).

4 Financial instruments for DRF

There are many literature references available on financial instruments used for DRF, most of which are published within this current decade. This shows the increasing attention DRF has received over the years. Many of the literature are reports published by international institutions such as the United Nations, through UNISDR, the World Bank group (2016b), European Commission (2017), and Asian Development Bank (ADB) (2013, 2017). Also, other private institutions have made important contributions to this area such as insurances companies (mainly Swiss Re and Munich Re), and other private research institutions, such as Institute European of Mediterranean (Fosse et al., 2018; Cambridge Institute for Sustainability Leadership (CISL), 2016; Center for Clean Air Policy, 2012). In addition, some authors have drafted reports on financial instruments on behalf of several institutions such as UNISDR (Suarez and Linnerooth-Bayer, 2011), UK Government (Linnerooth-Bayer et al., 2012), The Word Bank Group (World Bank, 2016b), Overseas Development Institute (ODI) (Watson et al., 2015; Tanner et al., 2015), and Inter American Bank (Miller and Keipi, 2005). Several countries have adopted diverse financial instruments considering their local circumstances, and type, frequency, intensity and impact of disasters, e.g., India (Gulati, 2013), China (He, 2015), Pakistan (Khan, 2015), Uganda (Nakalembe and Owor, 2016), Kenya (Ben and Byaruhanga, 2017), Sri Lanka (World Bank, 2016a), and the Philippines (Vidar and Medalla, 2015). In addition, many project specific cases studies have been analyzed by some researchers/institutions (Poundrik, 2011; King, 2005a; UNITAR/CAF, 2013).

Following the United Nation Climate Change (UNFCCC, n.d.), some of the main financial instruments commonly used for disaster resilience are grouped and briefly described in this section.

The most popular financial instruments are diverse modalities of catastrophe risk insurance which protects against low-probability, high cost events, and may take the form of micro and meso insurance bundling individuals' loans and insurance, catastrophe reserve funds, and insurance-linked securities. The main financial instrument in this group are: catastrophe risk insurance at national or regional level (with the possibility of including micro and meso insurance): Index-based insurance schemes; Group insurance. Examples of these instruments are: Agricultural Insurance Development Program, Caribbean Catastrophe Risk Insurance Facility, Climate Insurance Fund, Disaster Risk Financing and Insurance Program, Ethiopia Project on Interlinking Insurance with Credit for Agriculture, National Agricultural Insurance Scheme, Pacific Catastrophe Risk Insurance Pilot. Indeed, insurance is one of the main financial instruments (World Bank, 2016a; and other aforementioned references), that is crucial for addressing risk transfer (King, 2005b), especially for extreme and low frequent events, and dealing with many financial issues that could arise as a result of climate change related catastrophes (Gurenko, 2015; Linnerooth-Bayer and Mechler, 2005). This has been extensively applied in developed countries and has been the focus of attention in workshops conducted in many countries and regions, e.g., Asia (Lucas, 2015; Juswanto, 2017); and India (Skoufias and Bandyopadhyay, 2015).

Other relevant group of instruments are risk transfer and risk pooling schemes which help risk holders to spread losses widely across geographical areas, stakeholders or time, and may be considered at the micro, meso, national, and regional levels, and bundling risk in various forms enables risk holders to gain efficiency. The main financial instrument in this group are: tools to identify risks and appropriate responses: risk layering analysis, total climate risk approach; and other various financial instruments (credit, savings) linked to risk reduction measures. Examples of these instruments are: African Risk Capacity, Africa Disaster Risk Financing Program, Caribbean Catastrophe Risk Insurance Facility, Conditional Cash Transfer, Disaster Risk Financing Analytics, InsuResilience, Pacific Catastrophe Risk Insurance Pilot, Turkish Agricultural Insurance System.

Another important financial instrument is contingent funding that should be considered as a liability which may arise depending on the outcome of a specific event. This aims to cover the financing of immediate post-disaster liquidity needs, using soft trigger (contrary to reinsurance coverage) which has been subject of further research (Clarke and Mahul, 2011), to identify its optimal use as part of the sovereign risk financing strategy. Contingency finance is an alternative financial option, planned as early response and recovery measures. The main financial instrument in this group are: Contingency fund; Disaster relief fund; Preferential interest rate financing; Contingent credit; Microcredit capital to poor borrowers/small enterprises. Examples of these instruments are: Business Continuity Measures, Community-based Revolving Fund, Deferred Drawdown Options for Catastrophe Risk, Disaster Relief Fund, Stand-by Emergency Credit for Urgent Recovery,

Bond financing is another important mechanism to raise finance from the private sector. This tool enables governments to finance disaster relief and post disaster reconstruction without over-stressing the fiscal budget, allowing to move quickly with the necessary flexibility to attend an extraordinary demand for funds. The most suitable type of bond are Catastrophe bonds (cat bonds), which are high-yield debt instruments, usually insurance-linked to secure cash flow in case of a disaster. To facilitate the transaction, a special purpose vehicle (SPV) is commonly set up, which invests the money from investors to pay back interests, and the principal amount is returned if the disaster does not happen. Thus, these instruments allow the transfer of risk to bond investors. The main financial instrument are Catastrophe bonds and Ex-post bonds issued after a disaster to raise funds. Examples of these instruments are Caribbean Catastrophe Risk Insurance Facility and MultiCat Program. A specific type of bond are climate-themed bonds issued as fixed income, linked in some way to climate change solutions for, either mitigation or adaptation, to finance or re-finance risk reduction related projects. Although climate bonds are relatively new assets class, they are growing fast and there is currently a number of bonds with diverse type and range of climate alignment (e.g., low-carbon transport/building, water/waste management, clean energy, sustainable land use, etc.). In these cases, the issuing entity guarantees to repay the bond plus either a fixed or variable rate of return and investors can be institutional entities (e.g., pension funds) or individuals. The main financial instrument are Climate bonds and, Standard and certification schemes. Examples of these instruments are Climate Bond Standard, Resilience bonds and Green bonds.

Other innovative financial instruments are commonly referred as blended instruments, which combine grants with loans or equity from public and private financiers. These are usually presented as financing approaches to procuring financial resilience. Most innovative financial instruments are designed as ex-ante financing. Examples of these instruments are Attribution bonds, District Development Fund, Disaster Risk Financing Analytics single donor trust fund, Remote Sensing based information and Insurance for Crops in Emerging economies. Other more innovative instruments such as microfinance, social funds or catastrophe and regional pools (ProVentium, 2009) have been developed to aim for a more resilient society.

TABLE 5.1 Classification of main financial instruments according to different criteria.

Ex-post financing Budget contingencies Donor assistance (relief) Budget allocation Domestic credit External credit Donor assistance (reconstruction) Tax increase	**Traditional instruments** Solidarity Savings and credits Informal risk sharing Insurance mechanisms
Ex-ante financing Reserve funds Contingent debt Parametric insurance Catastrophe bonds Traditional insurance	**Innovative risk financing mechanisms** Index-based micro-insurance programs Public sector risk transfer National Insurance programs Catastrophe bonds Contingent credits Insuring donors that support governments Insurance pools among small states

(Source: Ghesquiere, F., Mahul, O., 2007. Sovereign Natural Disaster Insurance for Developing Countries: A Paradigm Shift in Catastrophe Risk Financing. The World Bank.; Linnerooth-Bayer, J., Hochrainer-Stigler, S., 2015. Financial instruments for disaster risk management and climate change adaptation. Clim. Chang. 133(1), 85–100.)

Table 5.1 below presents a summary of the frequently used financial instruments revealed during the systematic review. These are split, first, between ex-ante financing and ex-post financing. Historically, efforts have been made to mainly implement ex-post financing by means of public budget allocations and contingencies, donor assistance and credits; ex-ante financing has focused on planning of actions for DRR.

Features of financial instruments are necessarily linked to the level of risk transfer of DRR responsibilities and this has been analyzed in some researches (Linnerooth-Bayer and Mechler, 2007; Linnerooth-Bayer and Hochrainer-Stigler, 2015). An interesting research paper (Clarvis et al., 2015) links principles of resilience to specific financial instruments and mechanisms in order to explore how resilience principles can actually inform decisions around DRR and resilience building and investment focusing on longer-term systems perspectives. There has been some other work in the literature dedicated to index-based innovative financial instruments to provide risk coverage, most commonly known as parametric insurance, which set specific amount of compensations linked to different levels of risks (Chantarat, 2015; Chantarat et al., 2015; Baumgartner et al., 2017). There is a need for additional research identifying potential indicators that trigger financial instruments and mechanisms following the Sendai Framework (UNISDR, 2015), and there is currently an expert working group composed of scientific and academic organizations, civil sector, private sector and United Nations agencies, fostering these tasks.

5 PPP and disaster risk financing

The consideration of diverse forms of partnerships between the public sector and the private sector to cope with DRR issues have been included in some institutional reports and many researches have explored new potential mechanisms that may contribute to a higher resilience in the long term.

The term "Partnership" suggests a close relationship of two or more separate groups for the mutual benefit and the improvement of the overall environment in which they are operating. The mutual benefit of each partner should not be at the detriment of the other partner or partners. Thus, Public Private Partnerships (PPPs) refer to arrangements where the private sector mainly supplies infrastructure assets and services that traditionally have been provided by the government (IMF, 2004). According to Burger et al. (2008), a PPP is an agreement between a government and one or more private partners, by which the private partners provide a service. The service provision is made in such a manner that, the service delivery objectives of the government are aligned with the profit objectives of the private partners and where the effectiveness of the alignment depends on a sufficient transfer of risk to the private partners. The understanding of the partnership between the public and private sectors is further divested into elements of stakeholders and respective political context, society, market economics, and private industry (Agyemang, 2011).

While there is not a single definition of PPP, the essential elements included in this contract are: (1) a long-term agreement between a government entity and a private entity; (2) the provision of a new or existing public asset or/and service;

(3) the transfer of significant risks and management responsibilities, typically the financial one, to the private partner, and (4) the remuneration linked to the private partner performance (World Bank, 2017b). Therefore, not any form of collaboration between the public and the private sector may be considered a proper PPP. The adequate maintenance of basic and critical infrastructures while making society more resistant can contribute to reducing the risks and losses of natural disasters.

Several authors have proposed frameworks that use some form of public private partnership or collaboration to address DRM challenges under the basis that each party is responsible for the risk that can best manage in accordance to its capabilities. While some researches present a general framework (Busch and Givens, 2013; Abou-Bakr, 2012; Auzzir et al., 2014), other authors use project (De Groot et al., 2016; UNISDR, 2008; APEC, 2013) or country case studies (Bajracharya et al., 2012; Bajracharya and Hastings, 2015; Roeth, 2009; Mysiak and Pérez-Blanco, 2016; Van der Berg, 2015; ISDR. Asian Disaster Reduction Center, 2007; Swiss Re, 2011), to feature different schemes which identify potential frameworks.

In this research, a broad literature review process has been carried out and several key elements have been considered to identify and explain to disaster risk financing and inclusive DRM strategies. These strategies have emerged to involve the civil society together with the public authorities responsible for dealing and structuring actions to minimize the impact of natural disaster, what is considered Disaster Risk Reduction. Indeed, the main actors in the public sector are typically easy to be identified, though a number of new multilateral public institutions are getting involved in this area, such as the World Bank or The Asian Development Bank. However, it is in the private sector, where new key actors (whose tasks may be complementary), should be identified and included in a global and integral approach. In this regard, forms of partnerships with the private sector need to be explored and standardized, based on the public-private collaboration (PPC), agreements or partnerships (PPP).

The different forms of collaboration between the public and the private sector open a broad range for introducing innovations in the traditional financial instruments and mechanisms, especially from an ex-ante perspective, which enable better planning for DRM. Some financial instruments may be directly linked to specific indicators, which automatically trigger the coverage during disaster events. In this way, the society is more receptive to contribute to DRF, if actions such as disseminate knowledge, increase transparency and develop automatic triggers like early warning systems, are taken. Thus, the society participates in a global approach to cope with the climate change acquiring and investing in new instruments. Example of this consideration may be the issuance of green bonds aiming to raise money for the needed sustainable investments.

A framework approach for further research should include public and private actors, DR financial instruments and funding tools and forms of partnerships or agreements between the different actors, which enable risk transfer mechanisms. The framework should be completed with the identification of the main limitations and best practices to be implemented.

The public sector as the first responsible party, should promote the use of diverse type of grants through its different governments and institutions, both financial, providing funding sources, and non-financial, as regulation, trainings and capacity building for implementing a DRR strategy. The private sector, through the financial agents, banks, funds and institutional investors may provide private sources of funds but also cooperation and expertise through other acting agents, within different forms of participation and/or collaboration.

This framework should be considered as a tool for a dynamic process in which feedback from real experiences, generates a theoretical knowledge which will indicate limitations and best practices to follow for policy makers. The analysis of a number of cases studies may improve and reshape the conceptual framework.

The effective reduction and financing of catastrophic risks requires a combined response by both the public and the private sector players, especially for developing countries. Funds are scarce and have to compete with others resources aiming diverse activities against the global warming. In this context, PPPs may offer appropriate solutions to help governments dealing with catastrophes through financing. New considerations of PPP may contribute to improve the disaster preparation and adaptation. The private sector has the financial resources and has the capability to absorb many of the potential risks on a cost-efficient basis, as well as performing a broad geographical diversification to manage disaster risks. This risk management is provided by the insurance and reinsurance companies, which count with valuable experience and knowledge dealing with natural catastrophes. On the other side, the public sector has the key to set up the required institutional framework and regulation to enable effective partnership that complements and improves the limited public efforts and ensure financial response through an appropriate risk transfer.

Potential partnerships may explore different forms of risk transfer and risk sharing. Traditional private insurance may be enhanced and adapted for disaster risk needs by implementing public and semi-private insurance schemes, that keep rates considerably low to encourage people and corporations in the highly exposed areas to contract them. This may also improve risk prevention by compromising a better maintenance of the targeted facilities, which eventually become more resilient.

In this regard, insurance stimulates prevention and mitigation efforts if the pricing includes risk-adjustment clauses as an incentive to keep a proper conservation, set up protective measures and follow contract clauses to prevent additional risks

Box 5.1 Case study: The Spanish approach.

The Spanish "Consorcio General de Seguros" provides Extraordinary risks insurance (disaster risk insurance—when "force majeure" event takes place). However, this coverage requires acquiring some previous ordinary insurance policy coverage offered by the private insurance companies and includes a deductible amount, which is regulated by the government through the Ministry of Economy. The risk coverage aim to nature phenomena and include earthquake, tsunami, extraordinary floods, volcanic eruption, atypical cyclone and, meteorite and similar falling. It also includes, terrorism, rebellion, sedition or similar and damages caused by the army in peace times. Business interruption and continuity may be also included in a separate coverage.

The "Consorcio de Compensación de Seguros" (CCS), is an instrument that services the Spanish insurance sector. It is a public business organization that is attached to the Ministry of Economy, through the General Directorate for Insurance and Pension Funds. It performs many functions within the insurance field, and among which those related to (1) coverage of extraordinary risk, (2) combined agricultural insurance and liquidation of insurance companies stand out, and, (3) compulsory vehicle insurance. It has its own assets; independent of those belonging to the State, and its activity does not depend on any State budget.

which may be avoided. Governments may also act as reinsurers within the frame of national insurance programs in order to supplement private insurance schemes. These public actions should be limited and only be carried out to promote private insurance participation or contribution to reduce the individual cost of insurance but should not replace the reinsurance activity of the private sector to avoid reabsorbing back the risk transfer. An example is included in Box 5.1.

The framework approach could be considered a straw-man proposal to introduce a DRR compliance concept for further review in the future to more solid-sounding stone-man framework. In effect, another interesting approach will be to adapt the PPP framework to a resilience form. Disaster risk resilience (DRR) adaption for PPP has been developed in some countries, and then, a new form of PPP could be identified as DRR compliant.

Most countries have addressed the issue of DRR in PPP with the content and management of the Force Majeure clause in their contracting stage. Quite often, the main coverage of this clause is provided by the Government, acting as reinsurer of this type or risks through a state insurance company. Indeed, in many developed countries, the governments compensate for potential damages which may be associate to force majeure causes. Usually the risks included in force majeure clauses are well defined as well as the level of compensation. Sometimes, some of the risks may be considered "relief events" which are at the private partner's risk. However, in some emerging and less stable PPP markets, often risks are not clearly defined and specified and the force majeure clause management is not straightforward, especially if there are limited resources available to cope with disasters.

Where commonly there is not a specific resilience policy for public infrastructures, general legislation and the terms agreed in the PPP contracts regulate the sharing of the extraordinary risks which usually include disaster risks. In this context, especially important the considerations of force majeure clauses, both in their definition and scope as well as in sharing terms, which set the level of compensation from the public sector in a disaster event (World Bank, 2017a).

Another important issue is the insurance availability. Where natural risks are insurable, it is expected that the private partner acquires an insurance coverage and therefore the government or contracting authority will most likely be successful negotiating down a termination payment. When the risks are presumably uninsurable, either because there is not specific product available in the market, or for discouraging costs of the insurance, it is expected that government provides full compensation for the operation costs and possible termination payment. In emerging PPP markets, the compensation may have to be negotiated according to the priority guidelines and resources availability.

6 Concluding remarks

Considering the extensive literature on DRF and financial instruments, it may be concluded that during the last 3–5 years, a plethora of financial solutions to address specific disaster risks have been developed. However, with the exception of private insurance, most of them have been are government based financial instruments. This is a heavy burden for governments, especially when most of them have budgetary limitations, irrespective of whether they are developed or developing countries. Therefore, there is a need to involve private sector in an integrated manner for effective DRR solutions. Thus, new forms of public-private integrated financial instruments need to be developed based on the interests and roles of the public and private sector and considering their share of risks.

The proposed integrated approach between the private and public sector should be considered to optimize efforts in order to enhance the preparedness and recovery from hazards/disasters. The integrated effort should consider a stronger commitment and involvement of the identified key actors. The Table 5.2 proposes a summary of the different instruments, mechanisms and actors, considering the timing of the actions and the purpose of each stage of an integrated approach. This will be further researched in the future.

TABLE 5.2 Summary of instruments, mechanisms and actors by timing and stage.

Instruments/mechanisms	Ex-ante financing	Ex-post financing	
Actor/phase	**Emergency (short term)**	**Recovery (medium term)**	**Reconstruction (long term)**
Governments, regional cooperation, international public institutions, multilateral agencies	**Financial assistance (mostly ex-post):** – Emergency plans – Allocation for critical infrastructure – Government reinsurance programs – Uninsurable coverage – Budget allocations (e.g., grants) **Sovereign risk strategies** – Reserve funds – Contingent credit – Risk transfer mechanisms (e.g., Cat bonds, index-based, risk parametric)	**Compensations and financial assistance:** – Budget allocations (e.g., capital funds) – Budget contingencies – Domestic and external credit – Donations – Tax increase	
Private sector	Traditional insurance, risk parametric, index-based insurance or Cat bonds. Microcredit, donations	Private funds Donations	Private funds Reinsurance (Gov. support) Donations
Public private collaboration (communities, ONG, volunteers, private sector actors)	Emergency plan: emergency assistance	Evacuation and re-accommodation	
Public Private Partnerships (builders, operators, financiers, technical experts)	Critical infrastructure operation	Evacuation, re-accommodation, cleaning up, infrastructure operation and maintenance	Post disaster ad hoc PPP—for land regeneration, building, and infrastructure reconstruction

The previous table includes new forms of alliances involving private sector actors and civil society organizations within a global an integral approach which should be planned and coordinated by government leaderships. It is important to note that not all forms of interactions between the public and the private sector involve similar contractual arrangements. One particular form of collaboration is the Public Private Partnership which is featured by the existence of a long-term contract to produce a public asset or service, in which a significant part of the risk, especially the financing risk, is transferred to the private partner, who is remunerated in accordance to its performance. This form may be very suitable for many actions in the planning and recovery phase of disasters that are mostly related to infrastructure, which is identified as a field of further research.

References

Abou-Bakr, A.J., 2012. Managing Disasters Through Public–Private Partnerships. Georgetown University Press, ISBN: 9781589019515.

Agyemang, P., 2011. Effectiveness of Public Private Partnership in Infrastructure Projects. Master Paper, The University of Texas at Arlington, Arlington, TX.

Alfalla-Luque, R., Medina-Lopez, C., Dey, P.K., 2013. Supply chain integration framework using literature review. Prod. Plan. Control 24 (8–9), 800–817. https://doi.org/10.1080/09537287.2012.666870.

APEC, 2013. New Approaches on Public Private Partnerships for Disaster Resilience.

Asian Development Bank (ADB), 2013. Investing in Resilience. Ensuring a Disaster-Resistant Future., ISBN: 978-92-9092-949-9.

Asian Development Bank (ADB), 2017. Financing instruments and access to finance. In: Workshop on Building Resilience to Natural Disasters and Climate Change. ADB Presentation. April 2017.

Auzzir, Z.A., Haigh, R., Amaratunga, D., 2014. Public-private partnerships (PPP) in disaster management in developing countries: a conceptual framework. Procedia Econ. Financ. 18, 807–814.

Bajracharya, B., Hastings, P., 2015. Public-Private Partnership in Emergency and Disaster Management: Examples From the Queensland Floods 2010–2011.

Bajracharya, B., Hastings, P., Childs, I., McNamee, P., 2012. Public-private partnership in disaster management: a case study of the Gold Coast. Aust. J. Emerg. Manag. 27 (3), 27.

Baumgartner, D., Lobo, N., Carle, B., 2017. Disaster risk management and financing system, and corresponding method thereof. US Patent App. 15/438,261, 2017.

Ben, O., Byaruhanga, J., 2017. Post-disaster risk financing instruments as a strategic financing option for disaster risk reduction in the Kenyan National Disaster Platform. Innov. J. Bus. Manag. 6 (05).

Boudreau, L., 2015. Discipline and disasters: the political economy of Mexico's Sovereign Disaster Risk Financing Program. Fondation pour les études et recherches sur le développement international. Policy brief No. 128.

Burger, P., Bergvall, D., Jocobzone, S., An, D., 2008. Public Private Partnerships: In Pursuit of Risk Sharing and Value for Money. Organisation for Economic Co-operation and Development, Winterthur.

Busch, N.E., Givens, A.D., 2013. Achieving resilience in disaster management: the role of public-private partnerships. J. Strateg. Secur. 6 (2), 1.

Cambridge Institute for Sustainability Leadership (CISL), 2016. Investing for Resilience. University of Cambridge. https://www.cisl.cam.ac.uk.

Center for Clean Air Policy, 2012. Overview of NAMA Financial Mechanisms. www.ccap.org.

Chantarat, S., 2015. Index-Based Risk Financing and Development of Natural Disaster Insurance Programs in Developing Countries. PIER Discussion Papers, Puey Ungphakorn Institute for Economic Research.

Chantarat, S., Pannangpetch, K., Puttanapong, N., Rakwatin, P., Tanompongphandh, T., 2015. Index-based risk financing and development of natural disaster insurance programs in developing Asian countries. In: Resilience and Recovery in Asian Disasters. Springer, Tokyo, pp. 171–200.

Clarke, D., Mahul, O., 2011. Disaster Risk Financing and Contingent Credit: A Dynamic Analysis. The World Bank.

Clarke, D., De Janvry, A., Sadoulet, E., Skofias, E., 2015a. Disaster Risk Financing and Insurance: Issues and Results. ferdi.fr/World Bank.

Clarke, D., Mahul, O., Poulter, R., Teh, T.L., 2015b. FERDI–WB Disaster Risk Financing and Insurance Policy Brief.

Clarvis, M.H., Bohensky, E., Yarime, M., 2015. Can resilience thinking inform resilience investments? Learning from resilience principles for disaster risk reduction. Sustainability 7 (7), 9048–9066.

De Groot, K., Altamirano, M., Wiggers, A., 2016. Reducing the risk of water related disasters. DRR-Team Mission Report. Costa Rica. March 6–13, 2016. Kingdom of The Netherlands https://www.rvo.nl.

De Janvry, A., 2015. Quantifying Through Ex Post Assessments the Micro-Level Impacts of Sovereign Disaster Risk Financing and Insurance Programs. The World Bank.

European Commission, 2017. Financing a Sustainable European Economy. Interim Report, July 2017 By the High-Level Expert Group on Sustainable Finance.

Fosse, J., Petrick, K., Fiorucci, F., Moulet, M., Albarracín, J., Jubert, M.R., 2018. Green Finance in the Mediterranean. IEMed Policy study 3.

Ghesquiere, F., Mahul, O., 2007. Sovereign Natural Disaster Insurance for Developing Countries: A Paradigm Shift in Catastrophe Risk Financing. The World Bank.

Gulati, A.G., 2013. Financing Disaster Risk Reduction-the Indian Context.

Gurenko, E.N., 2015. Climate Change and Insurance: Disaster Risk Financing in Developing Countries. Routledge.

He, Q., 2015. Global climate governance and disaster risk financing: China's potential roadmap in transitional reform. In: UN Climate Conference in Paris.

IMF, 2004. Public-Private Partnerships. International Monetary Fund. Fiscal Affairs Department, Washington UK.

ISDR. Asian Disaster Reduction Center, 2007. Promoting Public Private Partnership in Disaster Risk Reduction Japanese Cases. www.adrc.asia/publications/psdrr/pdf/PPP-Finalized.pdf.

Juswanto, W., 2017. Promoting Disaster Risk Financing in Asia and the Pacific. adb.org.

Khan, M.A., 2015. Financing for disaster risk reduction in Pakistan. In: Disaster Risk Reduction Approaches in Pakistan. Springer, Tokyo, pp. 337–359.

King, R.O., 2005a. Hurricane Katrina: insurance losses and national capacities for financing disaster risk. In: CRS Report for Congress, p. RL22086.

King, R.O., 2005b. Hurricanes and Disaster Risk Financing through Insurance: Challenges and Policy Options. Congressional Information Service, Library of Congress, Washington, DC.

Ley-Borrás, R., Fox, B.D., 2015. Using Probabilistic Models to Appraise and Decide on Sovereign Disaster Risk Financing and Insurance. The World Bank.

Linnerooth-Bayer, J., Hochrainer-Stigler, S., 2015. Financial instruments for disaster risk management and climate change adaptation. Clim. Chang. 133 (1), 85–100.

Linnerooth-Bayer, J., Mechler, R., 2005. Disaster Risk Financing for Developing Countries. IIASA Presentation for the World Bank. IIASA, Washington, DC.

Linnerooth-Bayer, J., Mechler, R., 2007. Disaster safety nets for developing countries: extending public-private partnerships. Environ. Hazards 7 (1), 54–61.

Linnerooth-Bayer, J., Hochrainer-Stigler, S., Mechler, R., 2012. Mechanisms for Financing the Costs of Disasters.

Lucas, B., 2015. Disaster Risk Financing and Insurance in the Pacific. GSDRC Applied Knowledge Services.

Miller, S., Keipi, K., 2005. Strategies and Financial Instruments for Disaster Risk Management in Latin America and the Caribbean. Inter-American Development Bank.

Mysiak, J., Pérez-Blanco, C.D., 2016. Partnerships for disaster risk insurance in the EU. Nat. Hazards Earth Syst. Sci. 16 (11), 2403–2419.

Nakalembe, C.L., Owor, M., 2016, December. Satellite monitoring for early warning and triggering disaster risk financing in Uganda. In: AGU Fall Meeting Abstracts.

OECD, 2015. Disaster Risk Financing. A Global Survey of Practices and Challenges. https://www.oecd.org.

OECD/G20, 2012. G20—Disaster Risk Assessment and Risk Financing. A methodological Framework. A G20.

Ozaki, M., 2016. Disaster Risk Financing in Bangladesh. papers.ssrn.com.

Poundrik, S., 2011. Disaster Risk Financing: case studies. In: East Asia and the Pacific (EAP) Disaster Risk Management (DRM) Knowledge Notes Working Paper Series; no. 23. World Bank, Washington, DC.

Pro Ventium Consortium, 2009. ProVention—Practice Review on Innovations in Finance for Disaster Risk Management. https://www.preventionweb.net.

Roeth, H., 2009. Consultancy Project on the Development of a Public Private Partnership Framework and Action Plan for Disaster Risk Reduction (DRR) in East Asia.

Skoufias, E., Bandyopadhyay, S., 2015. Workshop/Atelier Disaster Risk Financing and Insurance (DRFI) Financement et Assurance des Risques de Désastres Naturels.

Suarez, P., Linnerooth-Bayer, J., 2011. Insurance-related instruments for disaster risk reduction. Global assessment report on disaster risk reducfion. UNISDR, Geneva.

Swiss Re, 2008. Disaster risk financing: reducing the burden on public budgets. Focus Report, Swiss Re, Zurich.

Swiss Re, 2011. Closing the financial gap: new partnerships between the public and private sectors to finance disaster risks.

Tanner, T., Lovell, E., Wilkinson, E., Chesquiere, F., Reid, R., Rajput, S., 2015. Why all Development Finance Should Be Risk-Informed.

Teh, T.L., 2014. Counterfactuals for the appraisal of disaster risk financing and insurance strategies. In: SDRFI Impact Appraisal Project.

Teh, T.-L., 2015. Sovereign disaster risk financing and insurance impact appraisal. Br. Actuar. J. 20 (2), 241–256.

Tranfield, D., Denyer, D., Smart, P., 2003. Towards a methodology for developing evidence-informed management knowledge by means of systematic review. Br. J. Manag. 14 (3), 207–222. https://doi.org/10.1111/1467-8551.00375.

UN Framework Convention Climate Change (UNFCCC), n.d. https://unfccc.int/topics/resilience/resources/financial-instruments#eq-7 Accessed 13 January 2019.

UNISDR, 2008. Private Sector Activities in Disaster Risk Reduction Good Practices and Lessons Learned. https://www.unisdr.org.

UNISDR, 2009. Terminology on Disaster Risk Reduction. https://www.unisdr.org.

UNISDR, 2015. Indicators to Monitor Global Targets of the Sendai Framework for Disaster Risk Reduction 2015–2030: A Technical Review. https://www.unisdr.org.

UNITAR/CAF, 2013. Fortalecimiento de la Resiliencia Ante los Desastres en America Latina. Intercambiando Conocimientos y Buenas Practices. https://unitar.org.

Van der Berg, A., 2015. Public-private partnerships in local disaster management: a panacea to all local disaster management ills? Potchefstroom Electron. Law J. 1727-3781. 18 (4), 994–1033.

Vidar, C.G., Medalla, E.M., 2015. Deepening Regional Cooperation for Disaster Recovery and Reconstruction: A Proposal for Proactive Approach to Risk Financing (No. 2015-21). PIDS Discussion Paper Series.

Watson, C., Caravani, A., Mitchell, T., Kellett, J., Peters, K., 2015. Finance for Reducing Disaster Risk: 10 Things to Know. Overseas Development Institute, Climate & Environment Programme, London. https://www.odi.org/sites/odi.org.uk/files/odi-assets/publications-opinion-files/9480.pdf.

Winspear, N., Musulin, R., Sharma, M., 2012. Earthquake catastrophe models in disaster response planning, risk mitigation and financing in developing countries in Asia. Geol. Soc. Lond., Spec. Publ. 361 (1), 139–150.

World Bank, 2016a. Fiscal Disaster Risk Assessment and Risk Financing Options: Sri Lanka. World Bank, Washington, DC. https://openknowledge.worldbank.org/handle/10986/24689. License: CC BY 3.0 IGO.

World Bank, 2016b. Disaster Risk and Environmental Resilience. World Bank, Washington, DC.

World Bank, 2017a. Resilient Infrastructure Public-Private Partnerships (PPPs): Contracts and Procurements. The Case of Japan. World Bank, Washington, DC.

World Bank, 2017b. Guidance on PPP Contractual Provisions. https://ppp.worldbank.org.

World Bank Group, 2014. Financial Protection Against Natural Disasters: An Operational Framework for Disaster Risk Financing and Insurance. WB, Washington, DC. https://openknowledge.worldbank.org/handle/10986/21725. License: CC BY 3.0 IGO.

Chapter 6

Resilience through flood memory—A comparison of the role of insurance and experience in flood resilience for households and businesses in England

Jessica Lamond[a] and Namrata Bhattacharya-Mis[b]

[a]*University of West of England, Bristol, United Kingdom,* [b]*University of Chester, Chester, United Kingdom*

1 Introduction

Resilience to disasters for households and businesses can be seen to be a function of underlying characteristics of a population and disaster specific adaptations or behaviors that address risk reduction at all stages of the disaster cycle (Cutter, 2016). Disaster specific adaptations for improved flood resilience include physical adaptation of urban infrastructure and buildings to limit flood damage and provision of resources for reinstatement through insurance or recovery grants (Jha et al., 2011; Kreibich et al., 2015).

In the United Kingdom the responsibility for risk management at a property level is held by the property owner. Funds for recovery are therefore not routinely provided by national or local government and individuals rely on insurance or other private funding sources. In this context the provision of adequate and affordable flood insurance has been a constant source of debate and negotiation between private insurers and the government, culminating in the launch of Flood Re in 2016 (Defra, 2013). However Flood Re does not provide coverage for small businesses or multiple occupancy property under leasehold arrangements that were previously included in the statement of principles (Association of British Insurers, 2008).

Authors have called for further understanding of the impacts of the advent of Flood Re on households and businesses at risk (Surminski and Eldridge, 2015). Furthermore government policy promotes greater uptake of property level flood resilience by property owners and occupiers as part of an integrated flood risk management strategy (Adaptation Sub Committee, 2012; Bonfield, 2016). The propensity of homeowners to adapt has been shown to be linked to flood experience but also to other attitudes and perceptions around the responsibilities of different stakeholders in reducing risk (Bubeck et al., 2012). Such perceptions are affected by the availability of insurance and government grants (Lamond et al., 2009; Harries, 2010).

Understanding of the needs and behaviors of owners of different types of property is therefore vitally important as the market for insurance and the provision of resources for reinstatement are rapidly evolving. Previous studies of households (Joseph et al., 2015; Lamond et al., 2009; Soetanto et al., 2017) and businesses (Bhattacharya-Mis and Lamond, 2014; Wedawatta et al., 2014) have measured issues related to insurance, response and recovery. In addition, recent UK government surveys have examined the availability of insurance for households and businesses nationally (see, for example, Dickman et al., 2015a,b); and Bhattacharya-Mis et al. (2015) combined surveys of households and businesses in two communities in Northern England. However, our study is unique in using a common sampling method and survey delivery to target businesses and households in frequently flooded locations across England. Moreover, this survey chose to focus on frequently flooded locations using a sophisticated sampling strategy that resulted in a sample of households and businesses with memory of flooding.

2 Methodology

The method of analysis and understanding was guided by an operational framework which sought to capture fuller understanding of the issues related to repeat flooding of both residential and commercial properties. The research used a survey

Strengthening Disaster Risk Governance to Manage Disaster Risk. https://doi.org/10.1016/B978-0-12-818750-0.00006-4

of frequently flooded locations in England to explore the different experiences and behaviors of households and businesses at risk from flooding with respect to insurance and recovery. Based on the characteristics of the study, the most appropriate data gathering instrument was deemed to be a survey of the sample population via postal questionnaire. The decision hinged on the type of information to be gathered and availability of resources within a limited period of time. The choice of postal questionnaires for this research, despite their acknowledged shortcoming of low response rates, was based on the need to:

- capture the large spatial distribution of data in different geographical locations with repeat flooding history;
- collect data from a large variety of respondents for both residential and commercial properties with diverse educational, social and economic background;
- provide freedom from social or peer pressure, ensure anonymity and promote honesty in responses;
- collect relatively large sample data at low cost.

To maximize the number of responses a large sample of 3000 properties each, was selected for households and businesses. To enhance response rate, follow up reminders were sent to those who did not respond to the first mailing. The surveys took place in 2015/16 shortly before the introduction of Flood Re.

Two different but overlapping sets of questionnaires were prepared for the residential and business properties. The major themes surrounding both questionnaire sets were: details of the flood affected property, the respondents' flood experience, their action and response for damage and future reduction, availability and affordability of flood insurance. The questions were mainly targeted to population who had experience of multiple flooding to their home or business; however data was also collected from people who were flooded once or had indirect flood experience. Space was provided within the questionnaire for open responses and additional information related to particular questions. Sample postcodes and full addresses were identified using a GIS platform by overlaying historical flood maps and Ordnance Survey property level address data. This selection method, while generating a high proportion of responses from flooded properties also resulted in a high level of undeliverable surveys, particularly to the non-domestic addressees as a result of out-of-date address details.

3 Results

Response to the survey was greater from households (17.4%) than from businesses (6.4%) but comparable to similar surveys of floodplain populations (Dickman et al., 2015a). A sample of 523 residential properties (294 flooded) and 172 commercial properties (110 flooded) were used for the analysis after cleaning and excluding substantially incomplete responses and non-deliverable postal returns. The inclusion of partially completed questionnaires maximized the use of information, however, as a result, there are differences in the sample response size for some questions as not all questions were answered by all respondents. Table 6.1 shows the summary statistics of the sample.

More residential properties (89%) are owner occupied than businesses (40.6%). With a high percentage of population in both residential and business properties living there over 20 years, the respondents are on average well established and, given the selection of areas that have flooded within the last 20 years, they are likely to be more aware of the local conditions in terms of risk of flooding than the general population. More commercial property occupiers had experienced flooding than residential occupiers, and the flooding was on average more severe. However more respondents among residential property sector were experienced in repeat flooding than that of commercial property respondents. While the severity of

TABLE 6.1 Summary statistics for the households and businesses (whole sample).

	Households %	Businesses %
Percent owning building	89.5	40.6
Long term occupation over 20 years Median length of stay	35.1 11–15 years	44.2 16–20 years
Percent experiencing at least one flood since 1995	52.4	63.5
Percent experiencing more than one flood since 1995	38.2	33.5
Percent experiencing severe flooding in their most recent flood. (Flooded households only.) Median damage level in their most recent flood. (Flooded households only.)	27.4 (major re-plastering and fitting) No damage	38.1 (major repair) Moderate damage

flooding for many (residential 68.9% and business 40%) was minimal (with no water entering their properties), a further large percentage (27.4% residential and 38.1% commercial) experienced severe damage requiring major re-plastering and refitting of fixtures and contents.

3.1 Access to insurance

Access to insurance in areas at risk has been measured and compared across flooded and non-flooded households and businesses. Table 6.2 provides a summary of the responses regarding access to insurance and various strategies undertaken to reduce the cost of insurance renewal.

Levels of insurance uptake are high in both households and businesses. There are very small differences in the proportion of households and businesses having insurance for their buildings and contents based on their experience of flooding. However the terms and exclusions for coverage show some variation based on type of property and flood experience. Households that have flooded in the past are most likely to experience higher premiums, flood exclusions and being effectively uninsured for flooding (16.7%) than non-flooded households (11.9%) and businesses flooded (8.2) or not (4.8). Accordingly they were more likely to adopt each of the suggested strategies to reduce the cost of coverage including the installation of measures to gain insurance. It is however interesting to see a higher percentage of businesses with no prior experience of flooding had seen a change in their premium as compared to residential properties without flood experience. Furthermore, businesses were much less likely to be subject to higher excesses for a flood claim. This acceptance of excesses can be a strategy adopted to reduce up-front premium costs by accepting some of the financial risk in the form of an amount deductible from any future claim. In terms of taking effective strategies in order to get affordable insurance, the common trend for both commercial and residential property owners is seen to be consulting with brokers, followed by calling several companies and using the internet to survey the market before future renewal. Installation of preventive measures as a strategy to reduce the cost of insurance is not seen as a popular option among either residents or businesses.

Business were also asked about the type of insurance they held and why (see Table 6.3). Some who had stated they had insurance gave no details of the type (some said they did not know as head office organized the insurance, others may have not included insurance that is held by other stakeholders). More businesses had contents insurance than buildings insurance, this could be because many businesses relied on others (e.g., their landlord) to insure the building. Owner occupiers were more likely to hold insurance than other tenures (except other). Unlike domestic policies businesses are making choices in their insurance based on what they need (for example one respondent said that they did not own the stock and therefore did not need to insure it) and what they can afford. New for old replacement is standard for domestic policies but indemnity

TABLE 6.2 Access to insurance across flooded and non-flooded properties (in areas at risk from flooding).

	Flooded households	Non flooded households	Flooded businesses	Non flooded businesses
Percent having either building or contents insurance	89.2	87.2	88.9	86.7
Percent having insurance policy that excludes flooding	9.1	1.5	N/A	N/A
Percent experiencing large changes in premium following flooding in the area	25.4	0.5	24.5	19.4
Percent accepting high flood excesses	39.1	12.9	8.2	4.8
Percent using broker	22.9	12.9	20.0	8.6
Percent phoning several companies	20.2	10.5	6.5	2.7
Percent using internet to find insurance	17.1	9.7	3.2	2.7
Percent asking neighbors/other business	10.3	5.5	3.2	3.6
Percent using adviser to find insurance	2.2	0.9	0.0	1.0
Percent installing measures to gain insurance	6.3	0.9	0.0	2.7
Percent effectively unable to insure[a]	16.7	11.9	4.5	3.2

[a] *Insurance refused/insurance without flood protection/unaffordable premium increase/excess so high I consider myself uninsured.*

TABLE 6.3 Type of insurance held by businesses by tenure.

Type of insurance held / Tenure	Buildings	Contents replacement	Contents indemnity	Contents any	Business interruption	Insurance any
Owned by self	76.8	44.9	46.4	72.5	40.6	79.7
Owned by company	58.2	56.4	32.7	65.5	40.0	74.5
Leased	47.6	50.0	28.6	64.3	40.5	73.8
Other	75.0	100.0	25.0	100.0	75.0	100.0
Total	63.5	51.2	37.1	68.8	41.2	77.1

(current value) insurance is common in business policies (37.1% having some of their contents covered in this way). Less than half of businesses held business interruption insurance, this could have implications for their financial viability after future floods.

3.2 Recovery strategies

Several recovery strategies were used by the respondents in order to recover faster from the impact of flooding. Focussing only on flooded households and businesses, this section examines the recovery experience of businesses and households for both insured and uninsured property. In terms of recovery experience, Table 6.4 shows that a high proportion of residential properties with experience of flooding were able to recover within a day as compared to businesses where the recovery took longer. However, a larger proportion of residential properties rather than business properties took more than 6 months to recover after flooding. Recovery time for households is highly correlated with level of damage as would be expected (rank correlation of 0.62 significant at 1%). Other reasons for differences in recovery time can include the type of property and nature of insurance coverage. On the other hand, the highest proportion of commercial properties both insured and uninsured took up to a month to recover, followed by 6 months and more. Severity of damage and time to recover were also correlated (0.55 significant at 1%). However, having insurance or not did not seem to affect recovery time.

While the residential properties mostly depended on insurance for their recovery the situation for businesses was slightly different. Although most held insurance (86%) other sources of recovery funding were also used. As Table 6.5 shows, less than 60% of business respondents, which were flooded at least once and have experience of recovery, chose insurance as their main source of funding.

TABLE 6.4 Recovery experience of insured and uninsured respondents across residential and businesses.

	% within flooded households (*N* = 278)	% within flooded businesses (*N* = 110)
Not relocated/recovered inside a day	65.5	5.5
Recovered/reoccupied in less than 1 month	2.9	58.1
Recovered reoccupied in less than 6 months	14.0	22.7
Recovered reoccupied in more than 6 months	17.6	13.6

TABLE 6.5 Main source of funding and help for business in recovery after the most recent flood.

(*N* = 104)	Normal running cost	Insurance	Savings/overdraft/loan	Contingency fund	Other
Funding for business recovery	23.1	57.7	14.4	3.8	1.0

A high proportion of businesses even after having previous flood experience still rely on savings and normal running costs for funding their recovery. Some have a contingency fund for such events but their proportion is very low. Fig. 6.1 shows the pattern of funding in relationship to the level of damage experienced. Those with small amounts of damage and loss are more likely to be able to (or to have to) absorb the loss without recourse to savings, loans or insurance (70%). Only 20% claimed. As levels of damage increase businesses rely more on insurance about 80 of the major and severely damaged used insurance. However a proportion also funded their own repairs from savings or loans and 20% of the most severely affected businesses had to use savings or a loan to make good after flooding. This points to a high vulnerability among a small group of severely damaged businesses some of which are not able to get insurance. However it also highlights the ability of businesses to recover in many cases without needing to claim on their insurance, especially for lower levels of damage and loss.

In their comments some respondents revealed that they were not able to get insurance, for example one said they had, "*History of property flooding. I have buildings and contents insurance but not flood cover. Insurance was unavailable after first flood in 1998*"; Or unaffordable premium: "*Cannot afford insurance premium. Premiums increased.*"

Another factor that was highlighted by business respondents is the higher excess charged on their insurance premium due to their location in the flood plain as one of the respondents mentioned: "*I have £5000 excess because I am in a floodplain.*"

Other reasons mentioned were lack of high value assets requiring insurance especially among small and medium size businesses, therefore no need was felt to insure against flood risk. In another case with a large corporate business, preference was given to self-insurance. This might in some way explain why despite having high access to insurance in general, some business owners use other sources of funding for their recovery.

3.3 Mitigation measures

Households and businesses were asked (in addition to the question of whether they installed measures to get insurance) whether they had installed mitigation measures at all. This was described in the questionnaire for households as "*making changes to your home or contents or buying protection measures to prevent damage to your property and contents.*" For businesses it was described as, "*adopting any preventative measures in your premises or the surrounding area.*" The results are show in Table 6.6. Within flood affected properties the number of households and businesses installing measures also varies significantly.

For residential properties 291 flood affected respondents responded to the question indicating whether they had made any changes for risk reduction. The responses show that 47.1% of total residential respondents had made changes, out of which 29.2% made some changes after the first flood. In their comments some residents gave reasons for not making changes such as not knowing what to do: "*Cos nothing has told us what to do to prevent ingress.*" Or on cost of measures: "*Cannot afford to have raised the floor and seal walls*". Other respondents answered that measures were unsuitable: "*Unfortunately as with so many properties PLP are not always suitable to protect properties effectively as in my property......Made enquiries and was advised ...the walls would not stand the external pressure.*"

Others were waiting for others to act: "*The change that need to be made is to solve the problem of the brook that caused the flooding. To build the bund that is being planned by the local authority. The site owners make all the changes not us*";

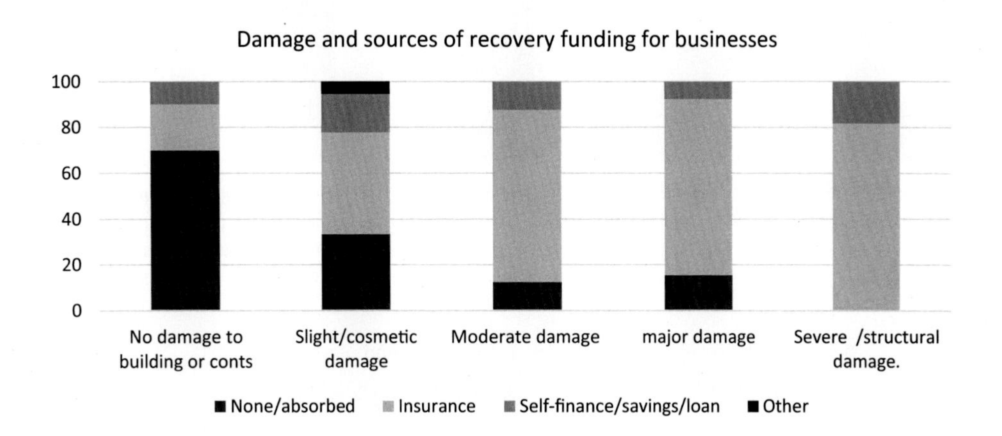

FIG. 6.1 Level of damage and sources of recovery funding for businesses.

TABLE 6.6 Respondents taking measures to reduce future flood risk across residential and business properties.

	% within households ($N=291$)	% within businesses ($N=106$)
Percent having made changes	47.1	66.4
Changes after first flood	29.2	50.0
Changes after second or subsequent flood	17.9	26.4

or being insured, "*damage to property is covered by building insurance covered by service charge*." Finally, some of the respondents said that low damage levels made expenditure on measures unattractive: "*Small expenditure taken from house-keeping. Only floods garden area which is several meters from house*."

Of the 176 households that had installed measures, 68.8% funded their measures using own savings with 8% borrowing to fund measures, 9.1% using insurance, and 14.2% using other sources.

Analysis of the experience before measures were installed (Fig. 6.2) shows that households were more likely to take measures if their damage was medium to major than if they were slight.

In the case of businesses, out of 106 respondents 66.4% mentioned that they made some changes or preventive measures for future risk reduction. As for households the majority made changes after the first flood. The percentage of businesses that made changes was higher than households after both first and subsequent experience of flooding. Many business respondents mentioned in their comments that there were further changes after the second flood even where changes were made after the first flood.

When asked about the measures they installed themes emerged such as avoidance for moving critical sites out of the flood plain for property owners some of the responses are as follows: "*After 2007 flood office moved across wharf to new office which is raised above flood level.... Storage of stock above flood level... Workshops still in flood level... Wharf platform higher now, due to relaying for new crane*"; "*Landlord has implemented measures; we have where possible moved all machinery away from flood areas*"; "*moved up to the second floor*"; "*moved critical items out of the way*."

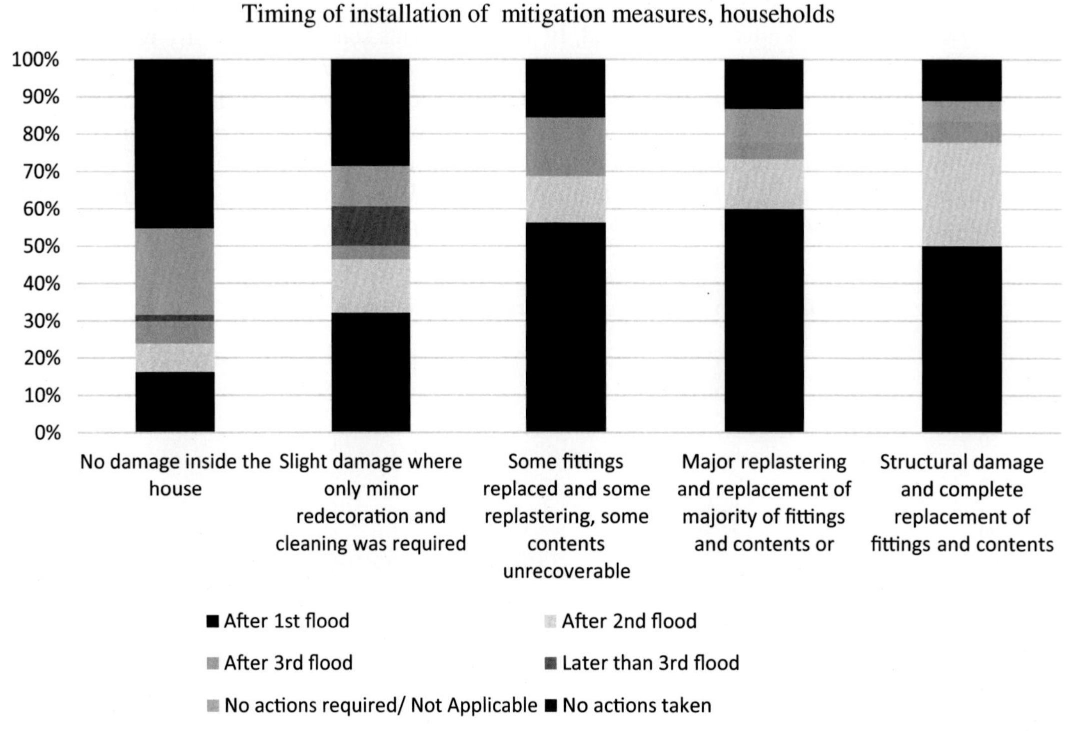

FIG. 6.2 Damage levels and experience before installing measures.

Other measures included installing barriers: "*Flood defences at front and rear of property. Also, a flood wall built inside of property to hopefully hold water out*"; "*Filled in vent in wall; sealed windows; flood gate.*"

Others implemented some recoverable measures: "*Wooden floor replaced by porcelain tiles; walls tanked & covered with easy to replace MDF panels; stud walls replaced by bricks; flood gate acquired*"; "*Tiled all downstairs floors. Furniture all modest, no antiques*"; "*changed flooring, removed skirting and put 'hard' flooring everywhere.*"

Other strategies ranged from signing up for the telephone alert system, flood plans, even sandbags despite their lack of effectiveness in many situations. Some also indicated they have flood action plans, however the numbers are very few. Some have mentioned that they have applied for government funds to support them to put up floodgates (information gathered from free text/additional information provided by informants).

3.4 Benefits of installing measures

There were also reported benefits in terms of damage reduction through implementing measures. About 16% of the residential respondents indicated that there was a marginal impact on damage reduction after taking up protection measures, 24% indicated they felt the measures significantly reduced their damage during the next flood and 26% said that they haven't had any damage caused during the most recent flood (Table 6.7).

This means that half of households making changes and suffering another flood had considerable or greater impact on damage. Conversely, about one third had seen no difference after making changes.

A variety of reasons for seeing no change or minimal change were also given for example higher flood levels than expected: "*First flood was nothing compared to second flood, so when we made changes we presumed if flooded again it would be similar. Unfortunately this was not to be. Lost everything as it happened during the night & early morning*"; "*Floodwaters exceeded barrier depth!*"

Or, having the wrong measures: "*Although the flood prevention barriers work well the damage to my property has been caused by surface water issues and the water has come up through floor to flood my property. This cannot be prevented by having the flood barriers in place*"; "*did not think that sandbags will make much difference. severe flooding will come up through floor*"; "*Water comes from different directions. Work on one side of the house - next time water comes from opposite side.*"

Other respondents mentioned not being present or able to install measures: "*The flood protection measures were taken after the first flood. The second flood - a flash flood - appeared virtually without warning and the protection measures were not in place, because I needed to be able to get to the cellar, sheds etc.*"

Other possible answers or related to not having much damage in the first place, "*Most of the damage in both floods was to the garage and gardens*" or that business occupiers were skeptical about some of their measures: "*Flood gate installed; does not work!*"; "*Door barriers in place but would be quite useless if flooded at 3m again!*"

4 Discussion

Since Pitt's (2008) review, engaging population at risk of flooding living in the flood plain is becoming more prevalent in the UK flood risk management scenario. As Priest et al. (2005) suggested in the context of flood insurance, the result of the perceived risk triangle is that human behavior is vastly affected by: knowledge and experience of flooding; awareness of risk; and expectation of flood management (Treby et al., 2006). The experience of businesses and their expectations for support are different from households, and Flood Re has sharpened the differences as small businesses are also excluded from coverage under the scheme.

Using a survey targeted at businesses and households in the same high risk areas has revealed differences in their experiences and strategies for recovery and mitigation. The results show high levels of availability of insurance but that there were more problems reported by homeowners than businesses in flooded communities in the lead up to the inception of Flood

TABLE 6.7 Residents indicating the reduction in damage after making changes in the next or latest flood.

	No changes	Marginal	Considerable to significant changes	No damage caused during next or most recent flood
Damage reduction in % (N=114)	33.1	16.1	24.0	26.8

Re. This may be related to the fact that businesses appear to be less likely to be reliant on insurance for their recovery and more likely to take steps to limit the damage to their premises and operations.

In comparison to other surveys the results are revealing. The percentage of households having insurance is high, with no difference between flooded and non-flooded households, this is not surprising in a UK context and it is at similar levels to other surveys such as Lamond et al. (2009) and Dickman et al. (2015a,b). The percentage of households experiencing difficulties that amount to compromised insurance is slightly higher than previously reported by Lamond et al. (2009), who saw 10% of those at moderate and significant risk and flooded having compromised insurance based on a 2007 survey of 5 recently flooded or significantly at risk areas. The number experiencing specific difficulties and pursuing strategies such as employing a broker are also higher in this population than in the 2007 survey and the differences between the flooded and non-flooded households is more marked. This may suggest that insurers are improving their risk assessments, perhaps based on increased availability of modeled data but also due to a larger number of claims since the 2000 floods. The percentage taking measures is also much higher in this sample, perhaps suggesting a growing awareness and willingness to put in place measures, but also perhaps due to the sampling strategy of areas flooded most frequently and the incremental effect of multiple flood incidents.

Exposure to risk, ineffectiveness of existing defenses, and uncertainty owing to climate change has given rise to the level of awareness among the risk affected population (Cripps, 2016) and the expectation is that the risk assessment should be done in a holistic manner at a catchment level while protection measures should be encouraged at property level. The establishment of the Flood Re scheme was to cater almost 350,000 households, but the exclusion of businesses and houses built after 2009 highlighted the obligation that new builds and businesses had to take in terms of installation of resistant and resilient measures. When comparing the results of this study with a similar study of the past post 2007 flood event it can be seen that there is an overall increase in the uptake of preparedness measures for businesses (Bhattacharya-Mis and Lamond, 2016). Uptake is more prevalent for businesses than households, which is reassuring given the advent of Flood Re. Despite the lack of government support there is a significant increase in use of business insurance as the source of recovery. From the results it appears that there is an increasing tendency to insure businesses to transfer greater risk to third parties in order to access capital for reconstruction for fast recovery. The level of acceptable risk depends on balancing business priorities one against the other to decide how much effort is appropriate to spend on hazard reduction. Businesses may not be inclined to take up actions to reduce environmental risk by investment of significant amount of money leading to financial stress. However, interestingly, despite their tendency to invest their own money in recovery, businesses are less likely to use higher excess to reduce their premium. For households it can be expected to see a reduction in the problems experienced in gaining insurance in the near future with the advent of Flood Re. The results may also indicate a shift in the tendency to adopt mitigation measures.

5 Conclusions

This study reveals differences between households and businesses in the way they use insurance and mitigation measures to limit the financial losses and disruption due to flooding. Businesses appear to be more self-sufficient than households and to be selective in their choice of insurance.

In the light of Flood Re and the absence of guaranteed cover for business this could represent an opportunity to encourage more businesses to invest in measures for low level events. It could also be possible to increase affordability for businesses through design of bespoke insurance products that recognizes their self-sufficiency through greater use of excesses or parametric type insurance.

Higher proportions of residential and business property owners and occupiers have reported taking action than in previous surveys. This may reflect some success in strategies designed to increase uptake that warrant further investigation.

The study is rare in estimating the damage reduction achieved through the implementation of measures by households. It is important to recognize that measures were on the whole effective in preventing some damage. Significant to total damage prevention was experienced by about half of those that installed them. Measures that did not prevent damage were sometimes ill-designed and did not address all the hazards a property might face. The importance of risk assessment and proper advice in the selection of mitigation measures is therefore evident. Better design of measures would improve the damage reduction potential.

Acknowledgments

We are grateful to EPSRC for funding the Flood MEMORY project and University of West of England to provide further funding for completion of the project.

References

Adaptation Sub Committee, 2012. Climate Change—Is the UK Preparing for Flooding and Water Scarcity? Committee for Climate Change, London.

Association of British Insurers, 2008. Revised Statement of Principles on the Provision of Flood Insurance. Association of British Insurers, London.

Bhattacharya-Mis, N., Lamond, J., 2014. An investigation of patterns of response and recovery among flood affected businesses in the UK: case study in Sheffield and Wakefield. In: Flood recovery innovation and response, vol. 184. WIT Transaction of Ecology and the Environment, https://doi.org/10.2495/FRIAR140141.

Bhattacharya-Mis, N., Lamond, J., 2016. Risk perception and vulnerability of value: a study in the context of commercial property sector. Int. J. Strateg. Prop. Manag. 20 (3), 252–264. https://doi.org/10.3846/1648715X.2016.1188174.

Bhattacharya-Mis, N., Joseph, R., Proverbs, D., Lamond, J., 2015. Grass-root preparedness against potential flood risk among residential and commercial property holders. Int. J. Disaster Resil. Built. Environ. 6, 44–56.

Bonfield, P., 2016. The Property Flood Resilience Action Plan. Defra, London.

Bubeck, P., Botzen, W.J.W., Aerts, J.C.J.H., 2012. A review of risk perceptions and other factors that influence flood mitigation behavior. Risk Anal. 32, 1481–1495.

Cripps, A., 2016. The bigger picture. RICS Building Control Journal (June/July, 7).

Cutter, S.L., 2016. The landscape of disaster resilience indicators in the USA. Nat. Hazards 80, 741–758.

Defra, 2013. Securing the Future Availability and Affordability of Home Insurance in Areas of Flood Risk. Defra, London.

Dickman, A., Langley, E., Silman, T., Harrold, P., 2015a. Affordability and availability of flood insurance (Report of the Defra Joint FCERM Research and Development Programme). Defra/Ipsos MORI, London.

Dickman, A., Langley, E., Silman, T., Harrold, P., 2015b. Affordability and availability of flood insurance: Findings from research with businesses (Final report FD2689). Defra/Ipsos MORI, London.

Harries, T., 2010. Household flood protection grants—the householder perspective. In: Defra and Environment Agency Flood and Coastal Risk Management Conference. Telford International Conference Centre, Telford, UK.

Jha, A., Lamond, J., Bloch, R., Bhattacharya, N., Lopez, A., Papachristodoulou, N., Bird, A., Proverbs, D., Davies, J., Barker, R., 2011. Five Feet High and Rising—Cities and Flooding in the 21st Century (Policy Research Working Paper 5648). The World Bank, Washington.

Joseph, R., Proverbs, D., Lamond, J., 2015. Homeowners' perception of the benefits of property level flood risk adaptation (PLFRA) measures: the case of the summer 2007 event in England. Int. J. Saf. Secur. Eng. 5, 251–265.

Kreibich, H., Bubeck, P., Van Vliet, M., De Moel, H., 2015. A review of damage-reducing measures to manage fluvial flood risks in a changing climate. Mitig. Adapt. Strateg. Glob. Chang. 20 (6), 967–989. https://doi.org/10.1007/s11027-014-9629-5.

Lamond, J., Proverbs, D., Hammond, F., 2009. Accessibility of flood risk insurance in the UK—confusion, competition and complacency. J. Risk Res. 12 (6), 825–840.

Pitt, M., 2008. The Pitt Review: Lessons Learned From the 2007 Floods. The Cabinet Office, London. http://archive.cabinetoffice.gov.uk/pittreview/thepittreview.html. (Accessed January 2019).

Priest, S.J., Clark, M.J., Treby, E.J., 2005. Flood insurance: the challenge of the uninsured. Area 37 (3), 295–302.

Soetanto, R., Mullins, A., Achour, N., 2017. The perceptions of social responsibility for community resilience to flooding: the impact of past experience, age, gender and ethnicity. Nat. Hazards 86 (3), 1105–1126. https://doi.org/10.1007/s11069-016-2732-z.

Surminski, S., Eldridge, J., 2015. Flood insurance in England: an assessment of the current and newly proposed insurance scheme in the context of rising flood risk. J. Flood Risk Manag. 10 (4), 415–435. https://doi.org/10.1111/jfr3.12127.

Treby, E.J., Clark, M.J., Priest, S.J., 2006. Confronting flood risk: implications for insurance and risk transfer. J. Environ. Manag. 81 (4), 351–359.

Wedawatta, G., Ingirige, B., Proverbs, D., 2014. Small businesses and flood impacts: the case of the 2009 flood event in Cockermouth. J. Flood Risk Manag. 7 (1), 42–53. https://doi.org/10.1111/jfr3.12031.

Chapter 7

Moving from response to recovery: What happens to coordination?

Emmanuel Raju
Global Health Section & Copenhagen Centre for Disaster Research, University of Copenhagen, Copenhagen, Denmark, African Centre for Disaster Studies, North-West University, Potchefstroom, South Africa

1 Introduction

Disaster risk management has seen the importance of coordination at various levels of planning and implementation, ranging from preparedness to disaster response, and in planning for long term recovery and sustainable development. These processes are not independent of one another. However, disaster recovery coordination has been a subject that has not been focused much on, and is considered to be one of the most under-researched areas of disasters (Smith and Wenger, 2007). Despite the unclear boundaries between disaster response and disaster recovery coordination, the present paper aims to identify if there are any differences observed in coordination from disaster response to disaster recovery. It intends to meet this aim by answering the following research question: Do stakeholders involved in various activities and efforts after the tsunami (2004) in Tamil Nadu, India express any differences in coordination between disaster response and recovery?

The study uses the case study of the Indian Ocean tsunami of 2004 that affected many countries in Asia. The data for this study is gathered from stakeholders involved in coordination activities in Tamil Nadu in India post-tsunami. There were many coordinating platforms that were set up after the tsunami. There were local-level coordinating platforms (usually referred to as district level) in each affected district. This stemmed from the experience of the massive Gujarat earthquake of 2001 in India. The civil society organizations in collaboration with the government saw the need to coordinate the many stakeholders coming with a variety of resources. The tsunami was one such case with massive media attention and immense international funding. Furthermore, there was a coordinating agency at the regional level in Tamil Nadu. The government agencies were actively involved in this process, although the coordination platforms were primarily managed by the civil society organizations.

2 Theoretical framework

There are many definitions of coordination in the context of disaster risk management. In a study of humanitarian relief chains, coordination is used to "describe the relationships and interactions among different actors operating within relief environment" (Balcik et al., 2010, p. 23). Coordination also means working beyond established structures, rules and procedures, which Suparamaniam and Dekker (Suparamaniam and Dekker, 2003) call "renegotiation." In an inter-organizational relationships study, Hall et al. (1977, p. 459) define coordination as "the extent to which organizations attempt to ensure that their activities take into account those of other organizations." As disaster situations bring together a number of actors, McEntire (2007, p. 291) suggests that coordination is "the harmonization of activities among diverse actors." The number and variety of actors make it difficult to cooperate and to coordinate, while broader consensus among them facilitates inter-organizational cooperation and coordination (Stephenson, 2005). For the purpose of this study, coordination is defined as "the act of managing interdependencies between activities performed to achieve a goal" (Malone and Crowston, 1990, p. 361).

One of the reasons for the complexities of disaster response arises from the fact that disaster situations bring multiple stakeholders together. Many actors ranging from public to private come into action (Granot, 1997; Katoch, 2006; Kory, 1998; Schneider, 1992), and there is even competition to be the first responding organization to the disaster (Stephenson, 2005). In large scale disasters, these actors are faced with huge challenges in disaster response and recovery. The organizations involved vary depending on the disaster and the context. Some of the prominent actors seen at different times during disaster response and recovery are government, non-governmental agencies, UN agencies and the affected

Strengthening Disaster Risk Governance to Manage Disaster Risk. https://doi.org/10.1016/B978-0-12-818750-0.00007-6

community itself. The government machinery and other organizations involved continue implementation of old tasks and/or create new roles or change existing roles to react and respond to the disaster (Quarentelli, 1994). Disasters, in other words, give rise to new ways of functioning depending on the local contexts and depending on the need for the situation (Schneider, 1992). Scholars have written about various characteristics of new ways of functioning during disaster response as "ad-hoc development of organizational structure" (Neal and Phillips, 1995, p. 330) and as "novel, non-traditional or non-routine" (Quarentelli, 1993, p. 74). Usually, governments are known to have more standard operating frameworks and to follow the standard procedures (Schneider, 1992). Further, the case of the tsunami highlights that both governmental and non-governmental organizations have formalized structures and also create structures that are ad-hoc and needs-based. The best examples are coordination structures that mushroomed in the wake of the tsunami in almost all affected countries. These ad-hoc and standard procedures during response and recovery raise critical issues for coordination in terms of operational procedures and responsibilities, guidelines and administrative issues. It is important to also note that stakeholders have different views on values, mandates and goals in disaster recovery (Raju, 2013b; Raju and Becker, 2013).

One of the key issues is that "the number and diversity of actors made coordination simultaneously more expensive and less effective" (Telford and Cosgrave, 2007, p. 11). What is crucial to the present chapter is the emphasis that there has been "little distinction between coordination at the operational level (who does what and where) and strategic coordination at the policy level (such as for joint advocacy)" (Telford and Cosgrave, 2007, p. 12). It should however be noted that coordination at policy levels is very different from the coordination at field level (Bennett et al., 2006). This may be seen in the way coordination structures are created after disasters. However, there is very little evidence of collaborative efforts between various stakeholders. There seems to be a disconnect between policy coordination and field level coordination. Telford and Cosgrave (2007, p. 12) attribute one of the major coordination constraints to the "absence of any agreed representative mechanism."

There is clear agreement on the importance of coordination. It is a highly held ideal to coordinate and cooperate for effective response (Drabek and McEntire, 2002; Granot, 1997; Quarantelli, 1997). The majority of the literature available in the field of disaster coordination relates to disaster response management, while disaster recovery is one of the most under-researched areas of disasters (Smith and Wenger, 2007), and needs more attention. It has also been observed that coordination was less challenging during the Indian Ocean tsunami response of 2004, as there was a common goal of ensuring basic services to the affected people, than during the long-term recovery and building sustainability (Christoplos, 2006). Another study concludes "despite unlimited trust in rapid reconstruction capacity, post-tsunami livelihood recovery has been chaotic and uncoordinated" (Régnier et al., 2008, p. 410). Similarly, a study of the Mozambique floods in 2000 highlights that "coordination worked better during the emergency period than during the recovery period" (Moore et al., 2003, p. 316), and states the need for further research in disaster recovery coordination (Raju, 2013a). Therefore, the present study is an exploration into the question if there are any differences identified in coordination during response and recovery.

3 Research methodology

To answer the research question, empirical data was collected through semi-structured interviews with experts of humanitarian organizations and state organizations who were involved in disaster response and recovery after the Tsunami of 2004 in Tamil Nadu, India. This required respondents who could give information about the early phases of the disaster and how coordination changed over time. Therefore, the respondents were identified based on their involvement with the tsunami from the response time, as well as their engagement and role in the coordination platforms at the district and state level in Tamil Nadu. Twenty-four interviews with fourteen respondents, conducted in February 2010 and March 2011, are used for this study. The respondents represented different organizations such as the government (local and regional levels), international NGOs, local NGOs, and civil society organizations. All the respondents held a leadership role in their respective organizations. Semi-structured interviews help identifying people's experiences in the process (Patton, 2002). Purposive-snowball sampling is used to select these respondents, which is deemed suitable to identify the respondents who would be able to provide the needed information (Silverman, 2010). Some of the respondents had switched organization and were in new job roles. The respondents identified had particularly worked in coordination agencies, with humanitarian organizations that were involved heavily in response and recovery in close connection with the coordinating platforms, the UN agencies and the government. Snowballing helped to select these respondents. However, the risk with snowballing is that it may lead to similar type of respondents with similar information. Case studies may be particularly useful with new research areas (Patton and Appelbaum, 2003). In this context, although it is not a completely new area, as highlighted already, the research areas are considered under-researched. The study does not claim for statistical generalizations, but as an attempt to a deeper understanding of coordination in response and recovery. Case study research is proved to provide analytical generalizations (Flyvbjerg, 2001). The effectiveness of a case depends on the selection of the case itself and the methods to investigate it (Yin, 2009). However, results may be transferred to other settings by "conscious reflection on similarities and differences between contextual features and historical factors" (Greenwood and Levin, 2007, p. 70). The interviews, lasting for approximately an hour, dealt with questions about the coordination during response and its transition to recovery, and what the differences were during the two time periods. The key ideas and notes were discussed with the respondents at the end of the interviews. These interviews were later transcribed and coded for the themes to emerge. The major themes were identified in relation to the research question about different activities relating to coordination during response and recovery. These themes are discussed in the sections below. Secondary sources such as policy documents, government texts, prior research conducted were used to support the analysis.

4 Empirical findings

All the respondents identified that there are differences in coordination between disaster response and disaster recovery although the boundaries are unclear. The findings are presented in the following sections as factors that respondents identified as different and affecting change in coordination patterns.

4.1 Stakeholders' mandates

One of the main factors that the majority of the respondents identify regarding differences in coordination during response and recovery is the changing nature of stakeholders. The respondents attribute this changing nature to the reasons of mandates and expertise of the respective organizations or government departments. The respondents state that there are more varied mandates and goals of stakeholders in disaster recovery, compared to the more unified mandate and goal of disaster response to establish basic services to the affected populations. According to one of the respondents' speaking in relation to the changing nature of stakeholders says that "very few organizations have a long-term sustainable agenda." The respondent also adds that it is usually the local organizations and organizations with a history of working in the region or who plan to initiate long-term projects that contribute to the sustainable thinking. The respondent continues to say that these organizations having short-term goals are a contradiction to the disaster recovery and risk reduction framework (keeping in mind that coordination is taking place for the goal of a sustainable disaster recovery). Varied mandates may be attributed to the sectoral coordination that recovery demands. While working with specific sectors, organizations tend to lose the overall picture of coordination for a larger goal.

The respondents from the district and state level coordinating agencies identify that, with differing mandates of organizations during response and recovery, they observe not only different organizations but also different personnel from the same organizations over time from response to recovery. According to the government authorities from the districts, they highlight that different departments are involved at varying points in time from response to recovery based on the need and the expertise required. In other words, the recovery process entails coordination with different organizations and government departments and with different personnel with varied experience and expertise. As identified, unlike response, recovery demands a more nuanced understanding of different functional sectors as well as the overall picture of recovery. Therefore, coordination in recovery is more challenging when compared to response. Having more short-term agendas, continuous changes in personnel in government departments and changing staff members in non-governmental organizations may contribute to reducing the speed and continuum of the recovery process.

4.2 Level of engagement

This factor refers to the engagement of stakeholders or organizations with the affected areas and communities. During the interviews, one-third of the respondents identify that coordination differs with the progression of time as a factor, i.e., "levels of engagement" (quoted by one of the respondents to the study) of responding organizations is identified as a major factor for coordination. This factor deals with the time invested in response and recovery activities. One of the interview respondents remarks that, "During response… there is much more engagement, everyday understanding of issues in different camps, constant follow-up. With logistics – who requires what? This has to be done on a more frequent manner." The same respondent also adds, "as we move from relief and response, the levels of engagement reduce."

All responding organizations (NGOs, state, civil society and the affected community itself) that are involved in long-term disaster recovery begin to engage in discussion and the shift moves from logistics-based to approach-based coordination (this is discussed as a separate factor later in the paper). Respondents also relate the reducing frequency of coordination meetings with moving into rehabilitation and long-term recovery. At this stage, during recovery, the engagement of any coordinating agency does not happen with all the NGO's at the same time or at the same level (as during response) whereas

NGO's and government departments along with UN agencies with specific areas of interest or sectors and working mandates start to come together. For instance, one of the respondents notes that, "coordination platforms will yield better quality results with quality inputs if it involves organizations with long-term experience and who have an intention for long term presence in the region."

4.3 Information

More than half of the respondents identify that there is a changing need for different types of information over time and also a changing nature of information coming in through the process of response and recovery. During the relief activities, some of the examples of the type of information required and processed for coordination activities between various stakeholders are related to registry of supplies, numbers of affected communities and beneficiary lists. This also includes keeping track of various organizations coming from outside to respond. As time moved from response to early recovery, one needed to know the exact number of houses damaged, the number of houses to be reconstructed, the number of boats lost, etc., where the differential aspect of sector-related/specific information came into play. More strategic type of information concerning policy making at a higher level was also needed. This is regarding aspects such as best practices, building codes and designs, designing recovery projects in livelihoods, education, etc. Respondents observe that, although it is very difficult and not necessary for all, it is important that government departments and other external agencies get access to relevant information for their work in time. Collecting, managing and disseminating this information itself became the core responsibility of some actors in the tsunami response and recovery. One of the staff members during the interview summarizes the functions of the state level coordinating agency "as coordination at different levels, information dissemination and formation of core groups. It also acted as a resource centre for information, trainings for staff members and policy advocacy."

Four of the respondents who were heavily engaged in information management, highlight that during disaster response, there is greater need for similar and general information (such as damage and needs assessments) to a bigger audience. In the tsunami, during relief, different actors from various organizations and the government had common access to information to enhance better coordination and reduce duplication of efforts. During coordination for disaster recovery, information was specific and directed toward a particular set of stakeholders. This difference, respondents attribute to the "sectoral" aspect of coordination. Here as one of the respondent notes "there is need for more detailed information and detailed assessments" for recovery and also adds "when someone collects information, different sets of information caters to different groups." The coordinating agency was used as a platform for sharing information and dissemination about affected field level realities ensuring a multi-stakeholder involvement in disaster response and recovery.

Recovery is in itself a complex process as it is highly contextual. Further, with regard to information, there is greater need for collecting and collating information from different stakeholders. Unlike response, as identified earlier, the goals change during recovery. Therefore, stakeholders collect information that may be relevant to their specific work and this marks a decline in sharing resources for working together in recovery. Information needs may be similar during response and recovery. However, during recovery deliberations with the communities about relocations, livelihood strategies, etc., need further studies.

4.4 Objective of coordination

The majority of the respondents working within the coordination platforms highlighted that mainly there are two types of coordination: (1) logistics-based coordination that takes precedence during disaster response, and (2) approach-based coordination that takes precedence during recovery. According to these respondents, during response the attention to logistics-based coordination involves coordination between the hundreds of organizations concerning the management of their logistics in terms of relief material flowing within the country and from abroad. With a more common goal of a unified response, respondents indicate that coordination is dealing with logistical issues of keeping track of the damages and needs of the affected population, maintaining a record of the organizations coming in, etc.

During disaster recovery, the majority of the respondents note that the deliberations and sector related coordination meetings (education, livelihoods, health, etc.) are directed toward building common strategies within the various sectors. During disaster recovery, the more long-term activities are in focus. For example, adopting housing designs or community-based livelihood interventions required more detailing of policies at the government level and more operational details for the implementing agencies. According to one of the respondents, during recovery, as the organizations are diverse in expertise and sector-specific goals, there is need for a discussion to adopt a common recovery strategy that must be tailor-made to suit the local context. This building of common strategies and a unified recovery plan (though contextualized according to the district and the sector) in the words of the respondent is "approach-based" coordination. This, according to the respondent, enables for planning and enhancing interventions while keeping in mind the common long-term recovery goal. The respondent also adds that as the organizations bring previous disaster-related experiences, it is necessary that their working paradigms are synergic for effective disaster recovery. The vast majority of the respondents note that synergy is required as the above-mentioned differences between stakeholders bring a range of strategies and intervention practices in the various sectors. The role of the coordination platform in this regard ensured that these strategies are discussed and suited not only to the context but also to the larger goal envisaged. One of the respondents also states that it is difficult to coordinate for a consensus of approaches as social processes are complex. This complexity delays the process of decision while coordinating for disaster recovery. In such situations, various actors have different perspectives of various approaches which may work or result differently than expected in long-term recovery. The difference therefore lies in meeting emergent needs in a timely manner in response versus facilitating a dialogue with the affected communities (which begins during response and escalates in recovery), deliberating on long term sustainable strategies for disaster recovery.

5 Discussion

The findings from the Tamil Nadu experience, presented under the four themes above, highlight differences in coordination during disaster response and recovery. These themes are not independent of each other but are mutually interdependent.

In a mega disaster like the tsunami, with a multitude of different stakeholders (Katoch, 2006; Telford and Cosgrave, 2007), coordination poses a huge challenge. The experience and expertise of each organization and government department contributes to the changing nature of stakeholders from response to recovery. However, it is already experienced that even when all these organizations come with rich expertise and experience, efforts to coordinate poses huge challenges (Granot, 1997). There is a quantitative and qualitative change in the stakeholders involved from response to recovery. This is also due to the fact that during disaster response the majority of the stakeholders have a common goal of providing basic services (Christoplos, 2006). However, our study reveals that there are changing goals due to sector related activities during recovery. This changes the personnel involved from response to recovery from various organizations according to their expertise and mandate. Therefore, it is not only the change in stakeholders that affects coordination but also the change in personnel involved. This change in the nature of stakeholders comes with the finding that different organizations bring with them different goals, mandates and expertise. Quantitatively, there is a clear difference in the number of organizations involved during response and recovery. The multitude of organizations that flew into the disaster site during response slowly decreased as recovery activities started taking prominence. Coordination is also affected by the stakeholders' time spent in the disaster affected region during response and recovery. Therefore, despite the variety of stakeholders involved at different times, the experience of organizations involved with local government and the affected communities from a long time and experience of working with disasters in the affected region and the country adds more value to the coordination platform for continuity and flow of coordination activities.

The second finding highlights differing levels of engagement during disaster response and recovery. During disaster response, coordination begins with a high intensity of engagement with the affected areas as there is need to understand issues in relief across the disaster affected region. The findings related to the level of engagement also identify a sense of urgency in coordination as one of the important factors for high engagement during relief activity. However, with progression of time and activities into early and long-term recovery, the matters of coordination become different with moving toward a discussion about sector-related issues between various stakeholders. Planning for recovery involves paying attention to more detail concerning important sectors, such as housing, education, livelihoods, etc.

Information needs as identified during the study are different during disaster response and recovery. One of the key challenges noted in literature relates to overload of information resulting in delayed activities (McEntire, 2007). During response, the information needs are mainly directed toward the common goals of providing services to the affected communities, whereas, during recovery, information is directed to cater to different sectors. The information generated and disseminated at the operational level during response focusses on the damages and needs of the affected region which may be of interest to the majority of the organizations involved during response. During the tsunami response, information was gathered from different sources and disseminated to various stakeholders as there is a common need for the same type of information. However, in the long term, during disaster recovery, it was difficult for a single coordination mechanism to sustain all functions relating to information due to many organizations and institutions being involved with many sectors or clusters that needed attention. This is also due to increasing needs for detailed information with the progression of time and activities in disaster recovery. During disaster recovery, as the finding highlights the need for "sectoral" information affecting coordination it is however important to coordinate for the common goal and unified recovery plan (Raju and Becker, 2013; Raju and Van Niekerk, 2013). It must also be clarified, that coordination platforms explain the need for acting as a bridge between the affected communities, responding organizations, the government and other stakeholders.

Also, coordination does not mean a top-down approach to consensus building but a platform to enable joint activities between stakeholders and for policy advocacy.

The nature of coordination identified two major approaches that take prominence during response and recovery respectively. This mainly relates to that one of coordinating logistics during response and the other as coordinating strategies during recovery. Although the factor of "nature of coordination" emerged from the interviews as a differential aspect of response and recovery coordination, it leads to a conceptual hypothesis. In the author's analysis, the first three factors discussed contribute to the differences in response and recovery. However, it is these three factors that build an argument to the "changing nature of coordination." This is more of a result of the interplay between various stakeholders being involved over time with changing levels of engagement and with changing information needs during response and recovery. Although, respondents note the difference in nature of coordination between logistics and approaches during response and recovery respectively, there is a very unclear boundary of the gradual transition from one form to the other. Further, this discussion also notes that the two types of coordination cannot be solely restricted to response or recovery operations. However, the difference lies in the focus and priority of adopting an approach-based coordination during disaster recovery. All the efforts during recovery coordination are modified to suit the local context which highlights the need for bringing together various approaches. In the long-term recovery, coordinating approaches involves looking at various thematic areas (such as education, livelihoods, shelter, etc.) that need attention and decide on interventions that suits the context best. As with many different perspectives and working patterns of the government and multiple organizations, an approach-based coordination indicates a common way of dealing with a similar problem. This helps in building an agreement for appropriate approaches keeping with larger recovery goal.

6 Conclusion

This study indicates that stakeholders involved in the tsunami (2004) response and recovery efforts in Tamil Nadu, India, express substantial differences in the coordination process between disaster response and recovery. The differences that respondents indicate are (1) the changing nature of stakeholders; (2) the level of engagement in activities; (3) information; and, (4) the nature of coordination itself. Although the study is limited in scope and number of respondents, it may contribute to the discussion and future research on the importance and scope of multi-organizational coordination in disaster recovery. Finally, the study contributes to initiate a discussion at a conceptual and operational level for "approach-based" disaster recovery coordination. Also, the study highlights that further contributions may be made by investigating into coordination and collaboration by linking disaster recovery with sustainable development and the sustainable development goals. The study highlights the potential for future research in identifying challenges in disaster recovery coordination.

References

Balcik, B., Beamon, B.M., Krejci, C.C., Muramatsu, K.M., Ramirez, M., 2010. Coordination in humanitarian relief chains: practices, challenges and opportunities. Int. J. Prod. Econ. 126 (1), 22–34. https://doi.org/10.1016/j.ijpe.2009.09.008.

Bennett, J., Bertrand, W., Harkin, C., Samarsinghe, S., Wickramatillake, H., 2006. Coordination of International Humanitarian Assistance in Tsunami-Affected Countries. Tsunami Evaluation Coalition, London.

Christoplos, I., 2006. Links Between Relief, Rehabilitation and Development in the Tsunami Response. Tsunami Evaluation Coalition, London.

Drabek, T., McEntire, D., 2002. Emergent phenomena and multiorganisational coordination in disasters: lessons from the research literature. Int. J. Mass Emerg. Disasters 20 (2), 197–224.

Flyvbjerg, B., 2001. Making Social Science Matter: Why Social Inquiry Fails and How It Can Succeed Again. Cambridge University Press, Cambridge.

Granot, H., 1997. Emergency inter-organizational relationships. Disaster Prev Manag 6 (5), 305–310.

Greenwood, D., Levin, M., 2007. Introduction to Action Research: Social Research for Social Change. Sage Publications, Thousand Oaks and London.

Hall, R.H., Clark, J.P., Giordano, P.C., Johnson, P.V., Van Roekel, M., 1977. Patterns of interorganizational relationships. Adm. Sci. Q. 22 (3), 457–474.

Katoch, A., 2006. The responders' cauldron: the uniqueness of international disaster response. J. Int. Aff. 59 (2), 153–172.

Kory, D., 1998. Coordinating intergovernmental policies on emergency management in a multi-centered metropolis. Int. J. Mass Emerg. Disasters 16 (1), 45–54.

Malone, T.W., Crowston, K., 1990. What is coordination theory and how can it help design cooperative work systems? In: Proceedings of the 1990 ACM Conference on Computer-Supported Cooperative Work—CSCW '90, pp. 357–370. https://doi.org/10.1145/99332.99367.

McEntire, D.A., 2007. Disaster Response and Recovery. Wiley, Hoboken, NJ.

Moore, S., Eng, E., Daniel, M., 2003. International NGOs and the role of network centrality in humanitarian aid operations: a case study of coordination during the 2000 Mozambique floods. Disasters 27 (4), 305–318. Retrieved from: http://www.ncbi.nlm.nih.gov/pubmed/14725089.

Neal, D.M., Phillips, B.D., 1995. Effective emergency management: reconsidering the bureaucratic approach. Disasters 19 (4), 327–337. Retrieved from: http://www.ncbi.nlm.nih.gov/pubmed/8564456.

Patton, M.Q., 2002. Qualitative Research & Evaluation Methods. Sage, Thousand Oaks.

Patton, E., Appelbaum, S.H., 2003. The case for case studies in management research. Manag. Res. News 26 (5), 60–71.
Quarantelli, E.L., 1997. Ten criteria for evaluating the management of community disasters. Disasters 21 (1), 39–56. Retrieved from: http://www.ncbi.nlm.nih.gov/pubmed/9086633.
Quarentelli, E.L., 1993. Community crises: an exploratory comparison of the characteristics and consequences of disasters and riots. J. Conting. Crisis Manag. 1 (2), 67–78.
Quarentelli, E.L., 1994. Emergent behaviors and groups in the crisis time periods of disaster, University of Delaware Disaster Research Center, Preliminary Paper No. 206. Retrieved from: http://dspace.udel.edu:8080/dspace/handle/19716/.
Raju, E., 2013a. Exploring Disaster Recovery Coordination. Lund University, Lund. Available from: https://lup.lub.lu.se/search/ws/files/4305738/4180232.pdf.
Raju, E., 2013b. Housing reconstruction in disaster recovery: a study of fishing communities post-tsunami in Chennai, India. PLoS Curr., 2004–2007. https://doi.org/10.1371/currents.dis.a4f34a96cb91aaffacd36f5ce7476a36.
Raju, E., Becker, P., 2013. Multi-organisational coordination for disaster recovery: the story of post-tsunami Tamil Nadu, India. Int. J. Disaster Risk Reduct. 4, 82–91. https://doi.org/10.1016/j.ijdrr.2013.02.004.
Raju, E., Van Niekerk, D., 2013. Intra-governmental coordination for sustainable disaster recovery: a case-study of the Eden District Municipality, South Africa. Int. J. Disast. Risk Reduct. 4, 92–99.
Régnier, P., Neri, B., Scuteri, S., Miniati, S., 2008. From emergency relief to livelihood recovery: lessons learned from post-tsunami experiences in Indonesia and India. Disaster Prev Manag 17 (3), 410–430. https://doi.org/10.1108/09653560810887329.
Schneider, K., 1992. Governmental response to disasters : the conflict between bureaucratic procedures and emergent norms. Public Adm. Rev. 52 (2), 135–145.
Silverman, D., 2010. Doing Qualitative Research, third ed. Sage Publications, London and Thousand Oaks.
Smith, G., Wenger, D., 2007. Sustainable disaster recovery: operationalizing an existing agenda. In: Rodriguez, H., Quarantelli, E.L., Dynes, R.R. (Eds.), Handbook of Disaster Research. Springer, New York, pp. 234–257.
Stephenson, M., 2005. Making humanitarian relief networks more effective: operational coordination, trust and sense making. Disasters 29 (4), 337–350. https://doi.org/10.1111/j.0361-3666.2005.00296.x.
Suparamaniam, N., Dekker, S., 2003. Paradoxes of power: the separation of knowledge and authority in international disaster relief work. Disaster Prev Manag 12 (4), 312–318. https://doi.org/10.1108/09653560310493123.
Telford, J., Cosgrave, J., 2007. The international humanitarian system and the 2004 Indian Ocean earthquake and tsunamis. Disasters 31 (1), 1–28. https://doi.org/10.1111/j.1467-7717.2007.00337.x.
Yin, R.K., 2009. Case Study Research. Design and Methods, fourth ed. Sage, London.

Chapter 8

Effects of regulatory frameworks in community resilience: Governance and governability in the Southeastern Pacific coast of Chile

Paula Villagra[a,b,c], Carolina Quintana[b], and Karla Figueroa[b]

[a]*Institute of Environmental Science and Evolution, Austral University of Chile, Valdivia, Chile,* [b]*Landscape and Urban Resilience Lab (PRULAB), Austral University of Chile, Valdivia, Chile,* [c]*Centre for Fire and Socioecosystem Resilience (FireSES), Valdivia, Chile*

1 Introduction

Disaster governance is a characteristic of community resilience—i.e. a set of processes that link and guide networked capacities to assure adaptation after disaster (Norris et al., 2008). Disaster governance refers to the interrelated set of institutions, instruments and norms aimed at reducing the impacts of a disaster (Tierney, 2012). In general terms, governance is a characteristic of an "interactive or socio-political government (…) where [agreements] between public and private actors prevail in order to solve social problems or create social opportunities" (Kooiman, 2005, p. 58). However, its effectiveness in developing countries has been highly contested. Several authors argue that attributes such as disparities in income, well-being, access to services, and political empowerment, as well as lack of resources, poor diversity of stakeholders, and missing links among established networks, hamper the effectiveness of disaster governance in developing countries, by reducing the interrelationship and communication among institutions (Tierney, 2012; Gall et al., 2014; Djalante, 2012). Consequently, disaster governance has been fragmented, deteriorating the capacity to coordinate after disaster in particular (Botero et al., 2015). Governance is especially criticized in this context since it implies that government is a necessary but insufficient steering agent, while the direction of society requires the capacities and resources of other social actors (Aguilar, 2013, pp. 298–299).

While in the Anglo-Saxon world the concept of governance still exists, the opposite happened in Latin America where the concept of governability has had greater attention, emerging as an alternative approach to governance. The word governability was coined for the first time in 1975 (Camou, 2011) during the development of a Trilateral Commission in which the emerging problems in the countries of developed capitalism were discussed, finding difficulties related mainly to the political stability of their governments and their competitions to meet the needs of the population. An evident imbalance was observed between the excessive number of social demands and the increasingly insufficient resources of the governments. This was found to be a real issue since for institutions such as the World Bank and the UNDP, governance is explained as "according to the ability of the government and the different social sectors to combine adequately […] satisfactory margins of equity, equal opportunities and social protection, and increasing degrees of citizen participation in political decisions" (Tomassini, 1996, p. 5). Within this context, governability emerged to tackle the gaps and the equilibrium between the different needs, demands and resources of all actors within a specific territorial space. Governability has evolved as a newer concept for encouraging integrated action among institutions, instruments and norms with the local inhabitants and the territory. Within the governability approach, an extensive collaboration is required to reach an efficient and safety management of the territory (Tierney, 2012). Indeed, all stakeholders should be involved and committed to cooperate and collaborate after disaster with similar expertise and at a variety of scales (Gall et al., 2014), including for example, developing different disaster relief functions.

Accordingly, governability has been proposed as a positive approach, that all nations would wish to achieve in order to assure a "state of dynamic equilibrium between the level of societal demands and the capacity of the political system (state/government)" (translated from Camou, 2011, p. 36). The idea of a dynamic equilibrium is particularly interesting in disaster

Strengthening Disaster Risk Governance to Manage Disaster Risk. https://doi.org/10.1016/B978-0-12-818750-0.00008-8

prone environments. It means that a system allows change, it includes "more space to move" because it has lower stability (Holling et al., 2001), which for the resilience thinking approach (Walker and Salt, 2006) is vital for adaptation after disaster. Indeed, the "resilience thinking" approach has become the "de facto" framework to improve preparation, response and recovery after disaster at the community level (Cutter et al., 2014). A system (e.g., community) that allows change fosters flexibility and redundancy in governability, which are relevant resilience attributes in a disaster scenario. Flexibility is a highly valued resilience aspect essential for governability (Gall et al., 2014) as it allows the adjustment of policies, regulations and functions. For instance, if one institution cannot play its role (i.e. evacuation), another could take it if the system is flexible enough to release resources and functions to a different organization. Redundancy is another resilience aspect that is much required in governability (Villagra and Quintana, 2017). A redundant governance approach to disaster assures that different stakeholders and organizations have similar roles. For example, an emergency plan can indicate that evacuation procedures could be developed by two or more organizations; then, if one organization fails, the other can take care of the evacuation process.

The Cuban response to Hurricane Sandy in 2012 provides an illustrative example of governability in time of disaster (Botero et al., 2015). The Civil Defense's flexibility to act together with the inhabitants of the city of Havana and the rest of Cuba prevented greater loss of human lives. The active participation of the population in the prevention plans and in the confrontation also gave redundancy to risk management in Cuba. When the government response was obstructed to act in the face of the disaster, the actions of prevention and confrontation of the coastal risk took shape at the local level. This was possible, since social organizations, institutions and companies in the local territory are legally trained and highly educated in matters of disaster response. In this way, they joined their capabilities, materials and human potential to respond to the emergency. The social approach has been key to achieving governance against coastal risks and hazards in Cuba, which has a civil defense system structured and organized at all levels of the country with active participation of the population. Cuba has a legal and regulatory framework that includes not only the Civil Defense, the institutions and agencies of the central government of the State in the prevention and minimization of damages and risks, but also its citizens. This is possible since everyone receives environmental and risk education and training in curricular and extracurricular activities. In Cuba 99.7% of the child population is enrolled in primary schools and 93.2% in the secondary level (Botero et al., 2015). In addition, the education and training of the population in relation to measures of safety, protection and prevention before, during and after disaster is periodic and multidimensional. For example, the Meteoro type exercise is held once a year throughout the country with the participation of the people, encouraging the simulation of the impact of different threats on the population and on economic institutions. As a result, the interrelationship reached between the local community and government and non-government institutions is highly dynamic, assuring the possibility of flexibility and redundancy in disaster response and recovery, resulting in a high degree of community resilience in case of disasters.

The complexity of understanding the extent to which the interrelationship among all institutions, instruments and norms is oriented to build a flexible and redundant governability approach to disaster that adds to community resilience has not been empirically studied as we know it. This chapter addresses this issue in the Chilean context. Particularly on the Chilean coast, it is important to explore whether the response to disaster tends toward governance or governability, and if this approach is flexible and redundant in times of disaster. After the 2010 earthquake and tsunami, the Chilean government made huge changes to the regulatory framework, particularly to emergency and territorial planning instruments (Herrmann, 2015). Nonetheless, the outcomes of these modifications have not been evaluated in terms of their impact on governability and resilience. Accordingly, the aim of this study is to analyze the extent to which the planning and emergency regulatory framework of the territory puts emphasis on generating a governance or governability approach that is flexible and redundant, in order to assure community resilience. For this purpose, the following three objectives are proposed: (1) Exploring the orientation of the regulatory framework to resilience, either positive or otherwise; (2) analyzing whether the regulatory framework fosters flexibility and redundancy in times of disaster; (3) and identifying the extent that the regulations established in the regulatory framework are in line with and includes the resilience capacity of the community. This is to find if the framework follows a governance or governability approach.

2 Methodology

We used a case study approach in the southern coast of Chile subjected to tsunami hazard, including the content analysis of the regulatory framework of the area to explore the orientation of the planning and emergency regulations and the extent whereby these assure flexibility and redundancy in a disaster scenario. Considering that the governability approach means the integrated action among institutions, instruments and norms with the local inhabitants, we used spatial analysis by contrasting the action and regulations of governmental and non-governmental institutions (coming from the regulatory framework), with the actions and needs of the community on the territory. The case study approach was particularly important in this respect. The study area included 14 coastal villages, some with urban, rural and indigenous character, distributed within 4 towns, 3 different communes, and 2 different regions. These areas have been previously used as case studies to explore the adaptive capacity of the territory, or community resiliency to disaster, with a focus on the needs, actions and resources used by the population in case of tsunami hazard (Fig. 8.1) (Villagra et al., 2017). Hence, this study provides useful information with respect to the resilience attributes and behaviors of the population that can be used for the coding process during the content analysis. It also provides information with respect to the resilience capacity of the territory from the point of view of the community, including its needs and actions, which can be contrasted to that evolving from the content analysis of the regulatory framework during the spatial analysis.

2.1 Content analysis

For the revision of the regulatory framework, 17 instruments were identified that regulate the case studies and promote their development through different measures and plans, established to manage emergencies in the event of a tsunami. These documents include: land use zonification; emergency plans; measures for prevention, response and recovery regarding the threat of tsunami; the role of the different emergency organizations; and the coordination of the different bodies after a disaster (Table 8.1). The 17 instruments were subjected to qualitative content analysis (Andreu, 2000; Krippendorff, 2004) using the Atlas-ti software v.8.0.41. The analysis was carried out in three steps: First, the 28 resilience indicators defined in previous studies of the area (See details in Villagra et al., 2017) were systematized to be used for the coding process. The indicators cover four resilience dimensions. The physical dimension includes aspects of urban morphology (e.g., density, number of open spaces). The environmental dimension includes the role of elements of the natural environment (e.g., coastal forests as buffers, wetlands as sources of water). The social dimension is linked to the characteristics of the population (e.g., levels of poverty, participation in emergency organizations), and the perceptual dimension relates to aspects that influence the response and actions of the community (e.g., risk perception, place attachment). Secondly, the presence or absence of these indicators in each instrument was reviewed, generating codes based on the frequency of mention according to the positive and/or negative orientation to resilience. For example, the indicator in relation to the amount of open space available for the response to the emergency in Mehuín had a frequency of mention 1, and with a negative orientation to resilience. This is because the planning instruments propose that the implementation of these spaces in Mehuín is within the tsunami inundation area, which is not useful for the post-tsunami community's ability to adapt. Third, semantic networks (Fig. 8.2) were created for each town and dimension, in order to represent the relationships found among indicators, dimensions and instruments, synthetically.

2.2 Spatial analysis

Due to the scant information found regarding the indicators of the social and perceptual dimensions in the planning instruments, and at the same time the difficulty of spatially representing this information, the overlap between the resilience capacity of the territory from the point of view of the community with that proposed by the regulatory framework was performed for the physical and environmental dimensions only. This information was spatially overlapped using the ArcGIS 10.5 software. First, all the information found in the documents was spatialized, generating maps of the physical and environmental dimensions. Data of previous studies (Villagra et al., 2017) was used for representing the orientation of the community (*community resilience maps*), and data of the content analysis of the regulatory framework was used for representing the orientation of planning and emergency instruments (*institutional resilience maps*). Subsequently, the application of the Clip tool allowed identifying the areas that overlap in the maps of the same dimensions, finding positive and negative overlaps, generating a *resilience capacity map* by town and dimension. Accordingly, a resilience capacity map in which the overlap is 100% positive indicates governability, while a resilience capacity map in which the overlap is 100% negative indicates governance. Intermediate percentages suggest tendencies toward one or the other approach.

3 Results

3.1 Orientation of the regulatory framework to community resilience

The semantic networks obtained during the content analysis, indicates the frequency of mention of the indicators for each instrument. Fig. 8.2 shows the results of Puerto Saavedra in detail. It shows that the physical, environmental, social and perceptual dimensions of resilience have 16, 4, 6 and 3 mentions in the instruments, respectively. From these, 21 instruments have a positive and 8 have a negative orientation to resilience. The semantic networks also provide information with respect

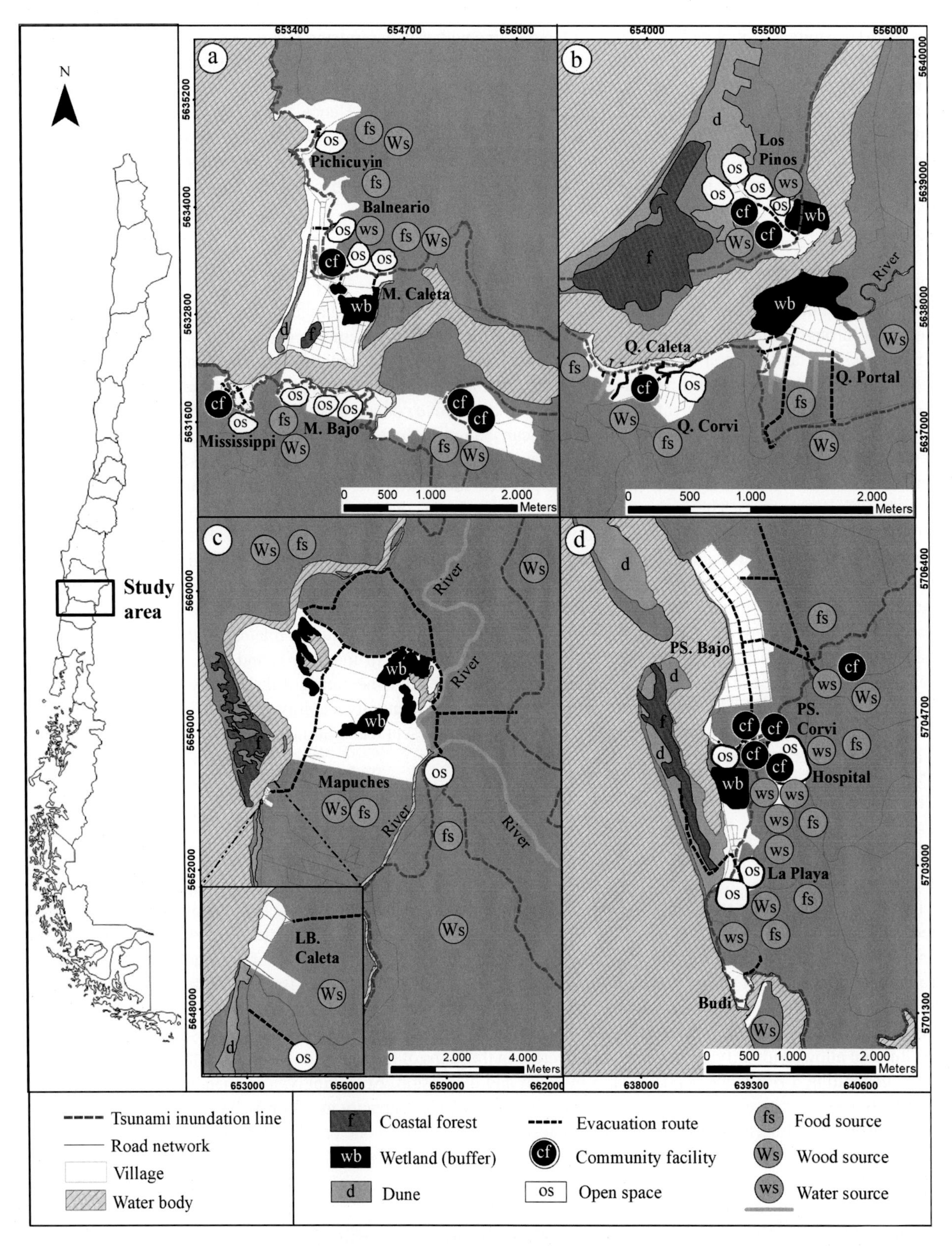

FIG. 8.1 Study cases. (A) Mehuín (ME), (B) Queule (QU), (C) La Barra (LB) and (D) Puerto Saavedra (PS). These maps show the resources of the territory people identify for satisfying their needs in case of tsunami hazard including: Open spaces useful for evacuation, for gathering and as for security areas; sites that provide food, water and wood (for heating); community facilities useful for temporal refuge; and natural buffer areas useful to reduce the tsunami effect. *(Source: Villagra, P., Herrmann, M.G., Quintana, C., Sepúlveda, R.D., 2017. Community resilience to tsunamis along the Southeastern Pacific: a multivariate approach incorporating physical, environmental, and social indicators. Nat. Hazards 88(2), 1087–1111.)*

TABLE 8.1 Regulatory framework under study.

Instrument	Study case	Scale	Type	Binding to	Description
1. Civil Protection Plan against Tsunami La Barra	LB	T	I	None	Includes the plans for the evacuation of the population to tsunami hazard, illustrating evacuation areas and routes, security zones and meeting points.
2. Civil Protection Plan against Tsunami Mehuin	ME	T	I	None	
3. Civil Protection Plan against Tsunami Queule	QU	T	I	None	
4. Civil Protection Plan against Tsunami Saavedra	PS	T	I	None	
5. Communal Development Plan Mariquina	ME	C	I	None	Policy and sectoral planning instrument at the community level with the objective of articulating and generating local and endogenous development initiatives, fostering the installment of social and economic development in a territorial development framework.
6. Communal Development Plan Saavedra	PS	C	I	None	Instrument for planning and management at the community level. Its purpose is to contribute to an efficient administration of the commune by promoting studies, programs and projects aimed at assuring the economic, social and cultural progress of the inhabitants of the Commune of Saavedra.
7. Communal Emergency Plan Mariquina	ME	C	I	None	Plan whose objective is to organize and coordinate with the actors of the municipal scope the handling of an adverse event, covering the emergency procedures during the stages of preparation and response.
8. Communal Regulatory Plan Mariquina	ME	C	N	None	Planning instrument that establishes land use, construction rules and roads requirements of the urban territories.
9. Communal Regulatory Plan Saavedra	PS	C	N	13[a]	
10. Communal Regulatory Plan Toltén	QU LB	C	N	None	Planning instrument that establishes land use, construction rules and roads requirements of the urban territories. It also rules human activity in risk areas.
11. Regional Emergency Plan Los Ríos	ME	R	I	None	Multi-sectoral plan that responds to emergencies, disasters and catastrophes, destined to the development of coordinated actions of the Regional System of Civil Protection.
12. Regional Plan of Civil Protection Araucanía	PS LB QU	R	I	None	Instrument for decentralized management with the objective of enhancing regional preventive capacities and bringing the management in emergencies and disasters to people, which aims to improve the quality of life of the inhabitants.
13. Regional Plan of Territorial Regulation Araucanía	PS	R	I	9[a]	Territorial planning instrument at regional scale, with the aim to support and manage public action, guiding public and private investments in the social, economic, infrastructure and physical-environmental fields. The objective is the equitable development of the Region and the physical distribution of space.
14. Risk Emergency Plan in case of Tsunami Queule	QU	T	I	None	Plan that provides information about the evacuation procedures in the event of Queule's tsunami alarm.

ME, QU, LB and PS stand for Mehuín, Queule, La Barra and Puerto Saavedra. The *Scale* refer to the area covered by the instrument (Regional (R), Communal (C) and Town (T). The *Type* indicates if the plan is normative (N) or indicative (I). The column with the head *Binding to* indicates the instruments that are linked to each other.

[a] *Indicates that both plans agree with respect to the management of dune systems for tsunami mitigation.*

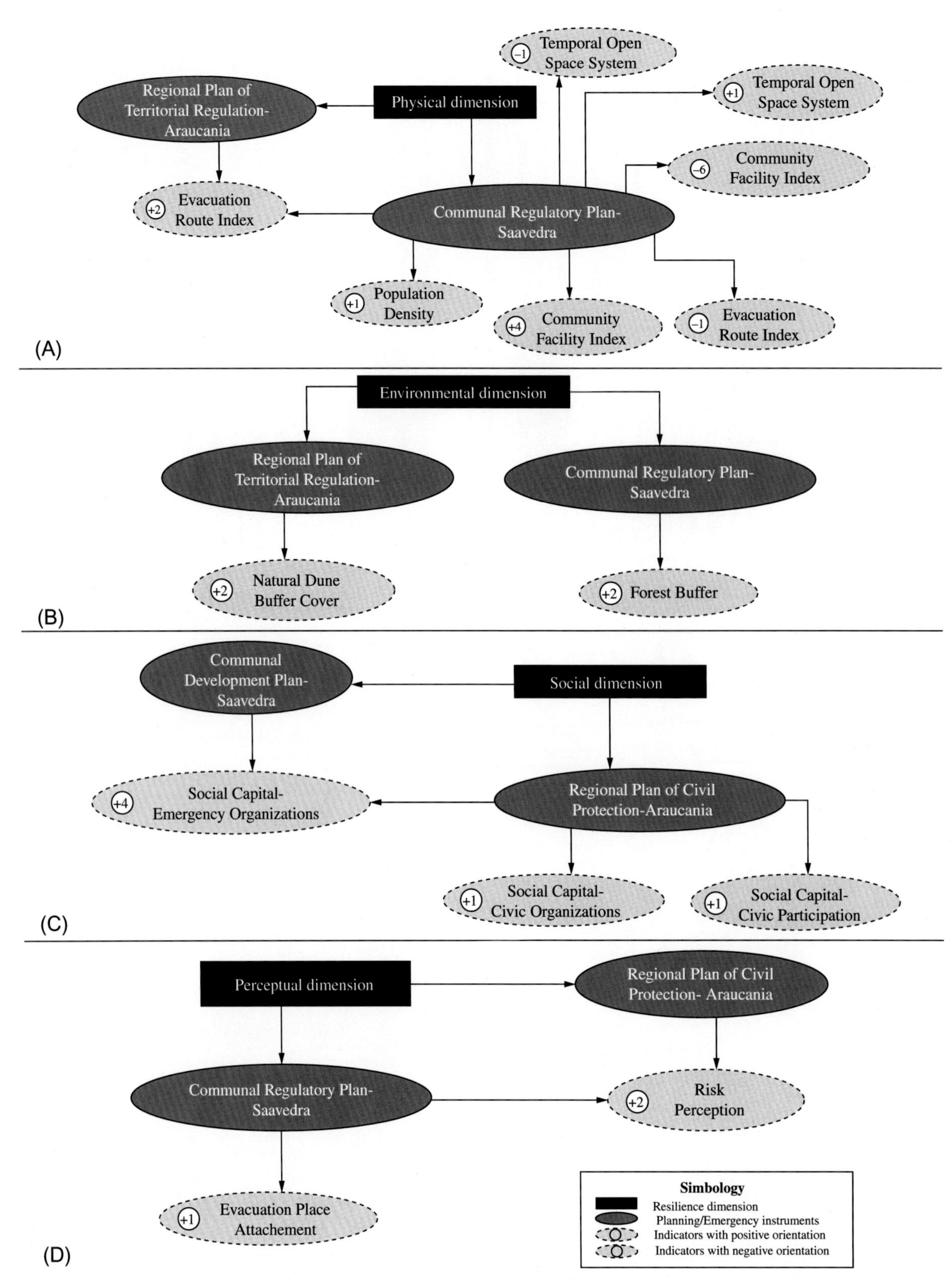

FIG. 8.2 Semantic networks—Puerto Saavedra. The figure shows the orientation of different instruments towards the dimensions of resilience. The *black rectangle* indicates the dimension; the *dark gray* ellipse indicates the instrument; and the *light gray* ellipse indicates the indicator of resilience to which it refers, positively (+) or negative (−). (A), (B), (C), and (D) show the instruments orientation to the physical, environmental, social, and perceptual dimensions, respectively.

to the instruments that refer to the same indicators. For instance, in Puerto Saavedra, The *Evacuation Route* indicator is addressed by the Communal Regulatory Plan and the Regional Plan of Territorial Regulation. The *Social Capital—Emergency Organizations* indicator is addressed by the Communal Development Plan and the Regional Plan of Civil Protection. In addition, the *Risk Perception* indicator is mentioned in the Regional Plan of Civil Protection and the Communal Regulatory Plan.

Particularly for the physical dimension in Puerto Saavedra (Fig. 8.2A), 8 positive and 8 negative mentions were found. The *Community Facility Index*, that refers to the amount of built infrastructure useful for emergency purposes, was found to have 4 positive and 6 negative mentions in the Communal Regulatory Plan. That is, the plan favors the allocation of schools, hospitals, police departments among other emergency infrastructure in the town; however, some of them are below the tsunami inundation line, so they are not useful in the case of a tsunami. In the same line, the plan favors the implementation of open spaces for refuge above and below the tsunami inundation line, which has an effect in the *Temporal Open Space System* indicator. This indicator refers to the set of open spaces useful for adaptation after tsunami and was found to have 1 positive and 1 negative mention in the plan. In addition, the Communal Regulatory Plan also suggests reducing *Population Density* (another indicator) in areas under tsunami hazard which favors resiliency. On the other hand, the same plan is only implementing one evacuation route, which was recorded as a negative aspect for the *Evacuation Route Index*, because redundancy is fundamental for resilience (Villagra and Quintana, 2017); hence, the more evacuation routes the more resiliency because if one route fails, another can take its function. The Regional Plan of Territorial Regulation was also found to have an effect in the physical dimension of resilience. In this case, 2 positive mentions were observed in respect to the *Evacuation Route Index*, that is, the plan suggests developing at least two evacuation routes for the area.

Within the environmental dimension of resilience of Puerto Saavedra, Fig. 8.2B indicates that the plans only refer to 2 indicators. One is the *Forest Buffer* indicator, that relates to the width of coastal forest for tsunami mitigation; the wider the forest the more mitigation (Forbes and Broadhead, 2007). Two positive mentions were found in this respect, because the Communal Regulatory Plan clearly establishes the conservation and management of forest areas located in between the town and the sea in two different sites. In the same line, the Regional Plan of Territorial Regulation encourages the protection of the dunes system that can contribute to tsunami mitigation as well. This was registered twice and relates to the *Natural Dune Buffer* indicator. The social dimension was addressed by two other plans (Fig. 8.2C). The Communal Development Plan addresses the *Social Capital—Emergency Organizations* indicator. This refers to the number of emergency institutions per capita in town, and a total of 4 positive mentions were coded in this respect. The plans encourage reinforcing the existing organizations in terms of emergency management and, in addition, indicates the creation of the Emergency Operations Committee (COE), which is a type of civil protection committee that is activated in times of emergency and is formed by emergency organizations and community groups. The Regional Plan for Civil Protection favors the *Social Capital—Civic Organizations* (i.e. the number of community groups per capita) and the *Social Capital—Civic Participation* (i.e. the percentage of the population participating in community groups) indicators. Finally, the perceptual dimension of resilience (Fig. 8.2D) was referred to in the Regional Plan of Civil Protection and the Communal Regulatory Plan. Both plans show a positive orientation with respect to the *Risk Perception* indicator, i.e. the extent that the community perceives tsunami risk. The instruments indicate that people must improve their perception of risk in their environment, and the resources they have to face it. They add that these parameters must be a substantive source to consider in the management of civil protection, which must be complemented with scientific and technical background, to establish a diagnostic basis for planning, according to local realities. Additionally, the Communal Regulatory Plan addresses the *Evacuation Place Attachment* indicator in a positive manner. The plan encourages improving the design and infrastructure of evacuation sites, so the community can use them during daily life, thus improving familiarity with these secured areas. The few studies about evacuation site choice suggest that people might prefer to evacuate to places where they find family members or friends, rather than public shelters (Wu et al., 2012). Hence, increasing attachment to evacuation sites may in turn improve intention of evacuation in case of tsunami as well as resiliency.

Overall, the comparative interpretation of the semantic networks of all towns indicates that there is a positive orientation trend to community resilience (Fig. 8.3A). References to resilience indicators with a positive orientation (PS 72.4%; LB 52.4%; QU 83.3%; ME 75%) are more numerous than those with a negative orientation in all case studies. Further insight provides the disaggregation of the data for each resilience dimension, after which we observed that the positive orientation prevails in all dimensions with the exception of the physical dimension. This means that indicators that refer to this dimension were more frequently found in the instruments with a negative orientation. As in the case of Puerto Saavedra, in the other towns the instruments propose the allocation of emergency infrastructure and open spaces for refuge under the tsunami inundation line, which is not supporting adaptation in case of tsunami. In addition, the environmental dimension is being considerably harmed for the case of Mehuín and La Barra in particular. The instruments propose here the reduction of coastal forests and the removal of sand dune in order to favor urban expansion. Hence, the capacity of the territory to mitigate the effects of a tsunami is reduced.

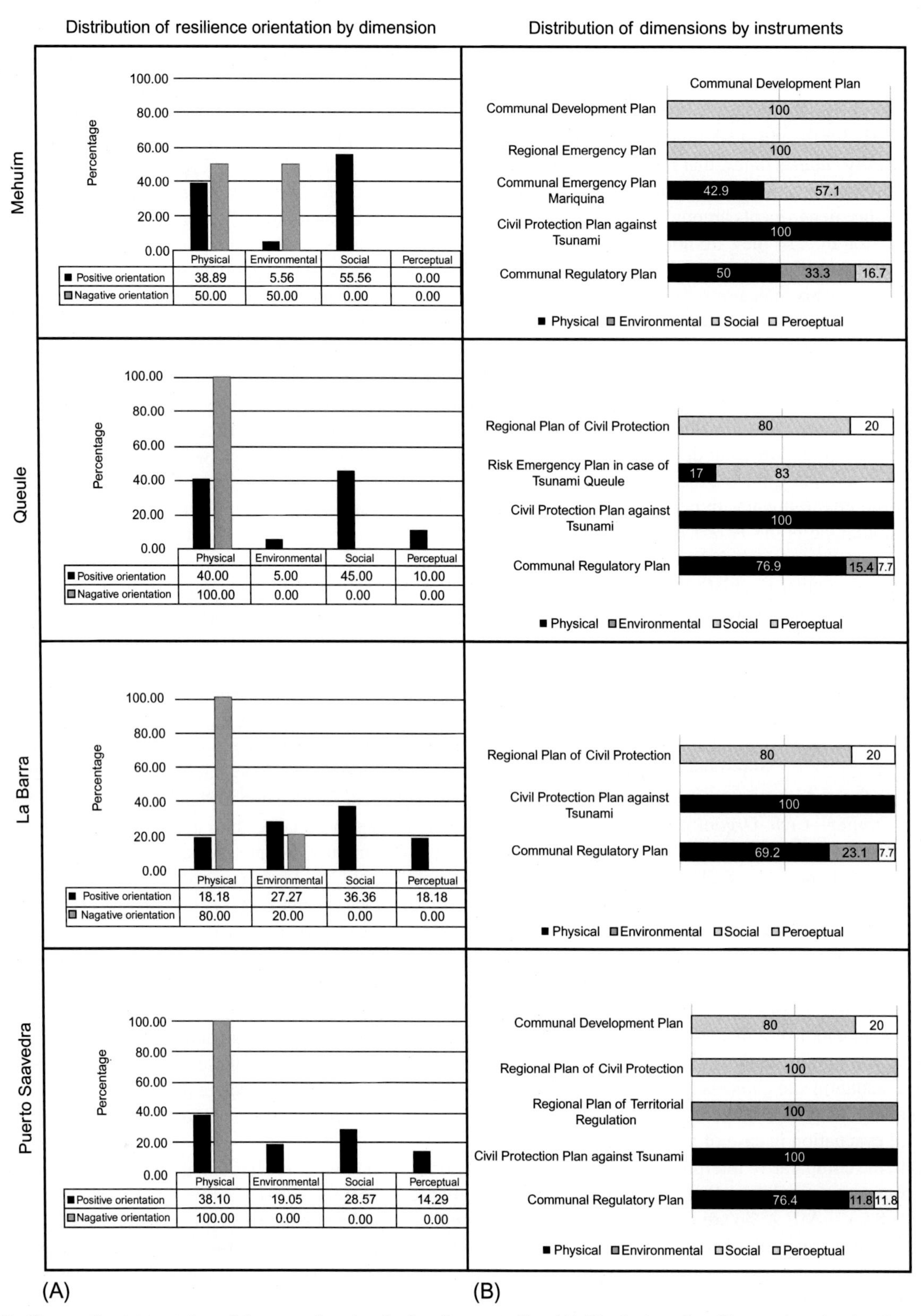

FIG. 8.3 Comparative interpretation of the semantic networks for all case studies. (A) Distribution of resilience orientation by dimension. (B) Distribution of dimensions by instruments.

With respect to the social dimension of resilience, in all case studies, the instruments coincide in promoting community participation in the different emergency agencies and in public and private institutions that participate in disaster risk management. This is positive for the social dimension since it fosters social capital in times of emergency. For instance in La Barra, the instruments increment the number of jobs vacancies for Firefighters, Red Cross, Civil Defense among others in emergency organizations. In Queule, the heads of schools are encourage to be educated in emergency procedures to collaborate during evacuation process and in Mehuín, it is suggested that ONEMI (National Emergency Office) provides advice to the communes for the organization during the tsunami alert and emergency period. Overall, the results with respect to the social dimension of resilience indicate that the focus of the instruments is the improvement of social capital; nonetheless, there are other important aspects not considered yet. For instance, regulations that address the *Population Poverty* and *Special Needs Population* indicators were not observed. It is relevant for resilience to address these aspects as an increment in poverty and the lack of solutions to evacuate people with special needs negatively affects the adaptive capacity of the community (Cutter et al., 2014).

Finally, indicators of the perceptual dimension were observed in a few number in comparison to the others (Fig. 8.3A: PS 14.29%, LB 18.18%, QU 10.00%, ME 0%), but were all positive. In all case studies, with the exception of Mehuín, the instruments encourage to improve *Risk Perception*. In Queule and La Barra for example, a study was already applied to the community to measure the perception of risk and the results are used in the instruments to establish regulations for the community. Indeed, it is noteworthy that the Communal Regulatory Plan that covers these towns emphasizes in fostering a culture of risk. Nonetheless, there are other perceptual resilience indicators not addressed yet. These include *Past Evacuation Experience* and *Objective and Subjective Knowledge*. Previous experience is generally positive on resilience as it leads to a higher risk perception (e.g., Trumbo et al., 2013) and to a sort of habit that makes those who have evacuated from previous hazards to be more likely to evacuate again (e.g., Murray-Tuite et al., 2012). Besides, objective knowledge, i.e., how much do people really know about the hazard and how to deal with it, together with subjective knowledge, i.e., how much do people think they know, are necessary conditions for resilient behavior (Stein et al., 2010). It is when people think they have the knowledge and the actual possibility to deal with the risk that they act resiliently and correctly perform coping behaviors (Demuth et al., 2016).

3.2 Flexibility and redundancy through the regulatory framework

The disaggregation of the data for each regulatory instrument provides further insight with respect to the extent that the regulatory framework fosters flexibility in times of disaster. The results plotted in Fig. 8.3B, indicate that the planning and emergency instruments are oriented toward one or two dimensions of resilience, with the exception of the Communal Regulatory Plan, since this plan rules the entire scope corresponding to the commune, not being specific to the emergency. This situation would not allow an institution to make decisions corresponding to another institution in the event that the latter fails after the disaster, since it would not have the corresponding capabilities or resources associated with the affected resilience dimension. For example, the same Fig. 8.3B indicates that in Puerto Saavedra, only the professionals of the Communal Regulatory Plan and the Regional Plan of Civil Protection have an influence on the perceptual dimension of resilience. Accordingly, if these professionals are not available after the emergency, there is no other team with the capacity to make decisions regarding the perceptual dimension of resilience, for example, to determine the location of an emergency shelter in relation to the perception of the risk of the community.

This limited coverage of the dimensions of resilience by each instrument together with the lack of linkage among them (Table 8.1) suggests little flexibility during the emergency. The lack of linkage implies that there is no interrelation between the planning and emergency instruments, which can negatively affect resiliency. For example, if an emergency plan establishes evacuation routes and planning instruments do not acknowledge them, the routes will not be built. When the plans are not linked, the decision of the planning instruments prevails since these are normative instruments, unlike emergency plans, which are indicative instrument, that is, they only suggest how the territory should develop. Indeed, Table 8.1 demonstrates that all emergency instruments are indicative and all territorial planning instruments are normative. Because instruments are not linked, the regulations on emergency response to disaster do not influence communal regulatory plans, reducing the territory's flexibility in the face of change.

Besides, the emergency instruments, and the civil protection plans in particular, provide information with respect to the role of institutions after disaster, which defines redundancy in times of disaster. The roles of emergency organization in times of disaster were identified during the content analysis of the regulatory framework. Eleven emergency activities were identified including; coordination, delivery of supplies (e.g., food, blankets), safe warding evacuation, the evaluation of the infrastructure, provision of information, provision of medical care, assuring security, searching for missing people, providing shelter, and the replacement of power supply and the water service. The frequency of mention of the distribution of these activities among emergency institutions after disaster indicates a low redundancy capacity in the governance structure. Evacuation is the only activity that can be addressed by various institutions (4 in Mehuín, Puerto Saavedra and Queule, and 3 in La Barra). In general, institutions that are part of every town such as the Municipality, the Police, Firefighters and the Sea Municipal Authority are trained to take care of the evacuation in case of a tsunami. Other activities such as the delivery of supplies and provision of information can be addressed by only two different institutions in PS, and the security and search of people can be addressed by two different organisms in Mehuín. The presence of the Maritime Authority in these two towns contributes to this result. The Maritime Authority has several other roles during a tsunami emergency such as that of reporting information, searching for missing persons, registering damage, and distributing supplies. However, this type of organism is only located in towns of higher communal hierarchies, such as Puerto Saavedra, which is the capital of the commune.

3.3 Effect of the regulatory framework on the resilience capacity

Data obtained in the content analysis was mapped for the physical and environmental dimensions of resilience. Fig. 8.4A illustrates the outcome of this process for the case of the physical dimension of Puerto Saavedra. This is an *institutional resilience map* for the physical dimension, indicating in black the areas with a negative orientation, and in white those with a positive orientation to resilience. The same type of map was prepared using the data of previous studies in relation to what the community finds useful for adaptation, and as a result, a *community resilience map* for the physical dimension of Puerto Saavedra was elaborated as well. The overlap of the *institutional resilience map* and the *community resilience map* for the physical dimension of Puerto Saavedra is illustrated in Fig. 8.4B. This is the *resilience capacity map*. It shows in white the areas in which both maps overlap with a positive orientation to resilience and in black shows the areas in which both maps overlap but the regulatory framework has a negative orientation to resilience. Fig. 8.4 shows that the areas in white include neighborhoods in which is compulsory to reduce urban density, and sites where it is indicated to implement evacuation routes, community facilities and green areas, that are useful for refuge and are above the tsunami inundation line. In black we find similar areas but in places that are under the tsunami inundation line or are in the process of being eliminated. Then, these areas are no longer useful for resilience.

The same process was developed for the environmental dimension of Puerto Saavedra and for the physical and environmental dimensions of the other case studies. As a result, we found different resilience capacity scenarios for each town. The spatial analysis for Puerto Saavedra indicates that the total area oriented to resilience after adding the institutional and community resilience maps is 39.77% of the study area. The institutional resilience maps cover in this case a larger percentage of the total study area (46.50%) than the community resilience maps (13.58%). It is positive to find in the results that 91.75% of the area that is covered by the institutional maps has a positive resilience orientation. In addition, the overlap between the maps with respect to the total study area oriented to resilience is 51.05%, with a positive overlap of 50.75%, suggesting a very low tendency toward governability. As it was mentioned earlier, a 100% positive overlap suggests governability because this result indicates that the orientation to resilience from the institutions and the community are in line, which is not the case here.

In the other case studies, the tendency is toward governance. Queule and Mehuín show a similar total area oriented to resilience with 48.35% and 43.29% respectively. However, in Mehuín the institutional maps cover a higher percentage of the total study area (31.92%) than in Queule (23.23%). In Queule, the area with a positive orientation to resilience covered by the institutional maps is slightly higher (97.93%) than in Mehuín (79.29%), suggesting that in both cases, most of the area regulated by the instruments is well oriented to resilience. A more negative aspect is that the overlap between the institutional and community maps is low, with 30.22% in Queule and 8.75% in Mehuín. Although in both cases the overlap with a positive orientation is higher than with a negative orientation, these results indicate that in most of the cases the view of the community with respect to the resilience capacity of the territory is dissociated from the view of the regulatory framework. This suggest a governance approach.

La Barra represents the worst-case scenario. First, the total area oriented to resilience is considerably lower than in the other cases with a 19.81%. Second, the institutional maps cover 1.51% of the total study area only, in contrast to the area covered by the community maps (18.31%). Indeed, and as a final point, the overlap between maps is 0.01%, from which 100% has negative orientation to resilience, meaning that none of the space the community finds useful for adaptation is being regulated, nor is it considered for adaptation in a disaster scenario by the regulatory framework of La Barra.

4 Toward governability and community resilience in the Chilean Coast

The results of this study reveal that the regulatory framework for the territory includes norms that affects resilience positively and otherwise. Even though the word resilience is not mentioned in any instrument, the reference to resilience dimensions and indicators was found in several of them as discussed earlier. The challenge here is to identify those that

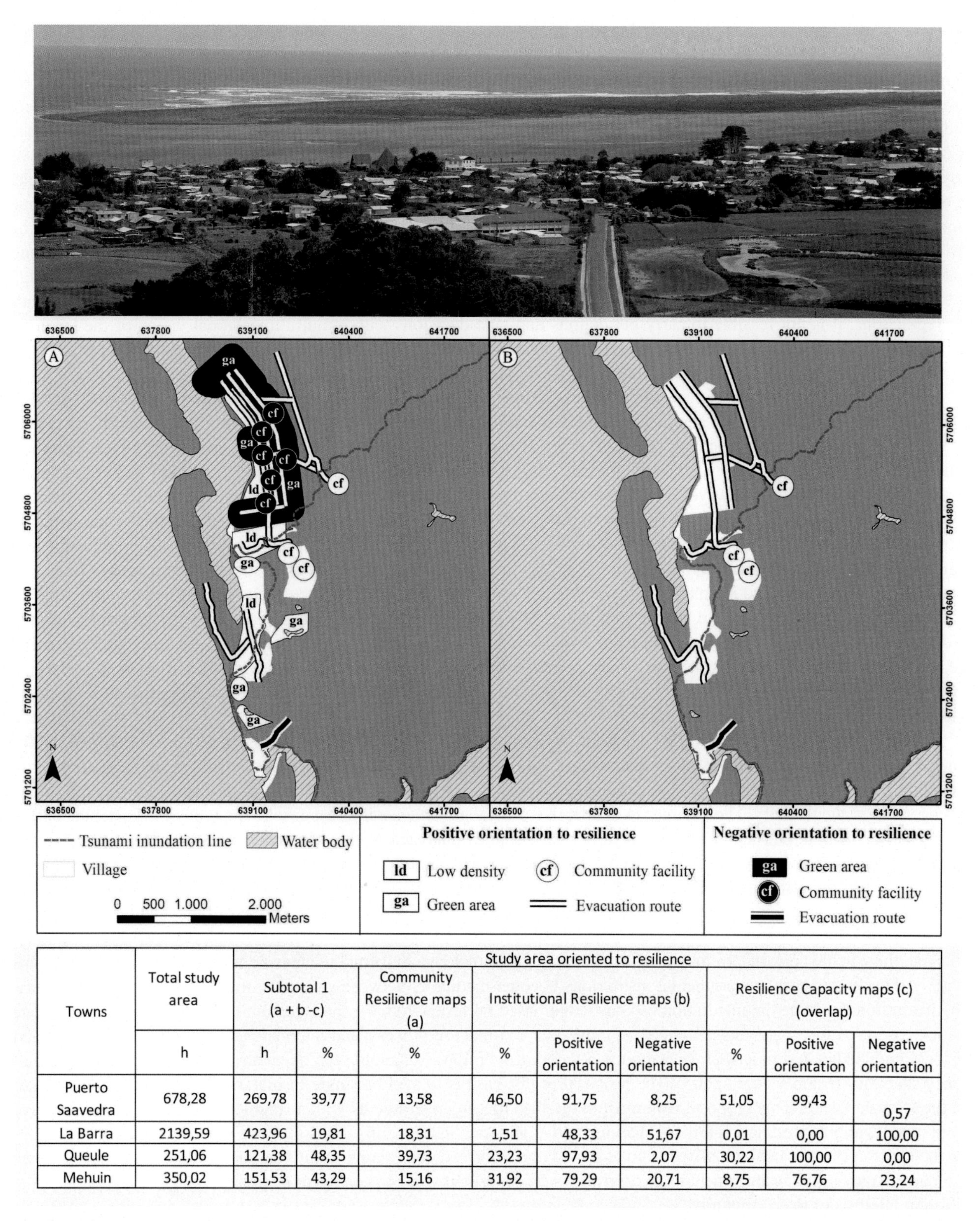

Towns	Total study area	Study area oriented to resilience								
		Subtotal 1 (a + b -c)		Community Resilience maps (a)	Institutional Resilience maps (b)			Resilience Capacity maps (c) (overlap)		
	h	h	%	%	%	Positive orientation	Negative orientation	%	Positive orientation	Negative orientation
Puerto Saavedra	678,28	269,78	39,77	13,58	46,50	91,75	8,25	51,05	99,43	0,57
La Barra	2139,59	423,96	19,81	18,31	1,51	48,33	51,67	0,01	0,00	100,00
Queule	251,06	121,38	48,35	39,73	23,23	97,93	2,07	30,22	100,00	0,00
Mehuin	350,02	151,53	43,29	15,16	31,92	79,29	20,71	8,75	76,76	23,24

FIG. 8.4 Resilience maps for Puerto Saavedra. The *institutional resilience map* (A) is to the left and the *resilience capacity map* (B) is to the right. The areas with a positive orientation to resilience are in white and those with a negative orientation are in black. The photo illustrates the view of Puerto Saavedra from one of the security sites 30 m.a.s.l. It shows the coastal dunes in between the Pacific Ocean and the town, the centric area of Puerto Saavedra at sea level, and the highway used as evacuation route. The table at the bottom is a comparative chart of the distribution of the areas oriented to resilience in each town. All percentages are calculated with respect to the *Total study area* with the exception of those under the *Resilience Capacity maps* that are calculated with respect to *Subtotal 1*. The percentages under *positive orientation* and *negative orientation* are calculated with respect to the area covered by the map they refer too.

have a negative orientation to resilience and modify them accordingly. This may be only possible if a "resilience thinking" approach (Walker and Salt, 2006) is undertaken by planners and managers. The "resilience thinking" approach, introduced from ecology to the human environment, has become the "de facto" framework to improve preparation, response, and recovery after disaster at the community level (Cutter et al., 2014). Nonetheless, the misleading understanding of the concept in the planning realm commonly misguides the elaboration and implementation of strategies meant to ensure the resilience of communities (Blanco et al., 2009), as is the case we found in this study for southern Chile. Resilience is the ability of a system to experience unexpected and extreme changes without crossing the threshold into a different system regime. In other words, it is the ability of a system to absorb disturbance and reorganize after change in order to retain the same function, structure, identity, and feedback capability (Walker et al., 2004). When resilience is applied in a community, the resilience thinking approach encourages actions that address the multidimensionality of resilience. In this study we found that each instrument tackles only a few of the resilience dimensions, i.e. physical, environmental, social and perceptual, with the physical and environmental dimensions being the most addressed. In addition, since the instruments are not binding regulations, it is not feasible to carry out actions and take decisions with a multidimensional vision of the territory. The multidimensionality of resilience should be reflected in the planning of communities if the aim is to build resiliency in disaster prone scenarios (Vale and Campanella, 2005; Cutter et al., 2014).

Another challenge that emerged from this study is to make the entire regulatory framework of the coastal areas in Chile more flexible, so the community can take a more relevant role in times of disaster, leading to a governability approach. As observed in the Cuban case study presented earlier (Botero et al., 2015), flexibility is possible if the experts in emergency and planning act together with the community, and this may also increase redundancy by having more alternatives to respond to disaster. This, however, is only possible with a well-educated and informed community in terms of disaster management as is the case in Cuba. In the case studies we explored in Chile, we can infer from the results of the semantic networks that the instruments are putting many resources in preparing and educating the community by the establishment of regulations that increases social capital. In particular, the aim is to increase the number of social and emergency organizations and of citizens involved in emergency procedures. This is a good beginning since social capital refers to the capacity of people to get together to respond to disaster positively (Walker and Salt, 2006), and this can be a very important resource in developing countries to complement or replace weak local governance structures. Many studies encourage reinforcing the role of the community in disaster governance because developing countries have been described as examples of the "messiness" of contemporary coastal governance, being a mixture of hierarchical, moderately collaborative, and somewhat integrated management (Evans et al., 2011).

Adaptive governance is an important approach to consider at this stage because it has been more addressed in the literature (e.g., Berkes and Folke, 1998; Brunner et al., 2005; Hahn et al., 2006), and has similarities with the governability approach, since both put emphasis in the equilibrium between experts and the community with respect to their role in the management of the territory. The emergence of adaptive coastal governance is a response to the type of intractable conflict among policy, people and nature, observed in the case studies analyzed in this chapter as well. When adaptive governance happens, society has reached the capacity to include environmental and social feedback into policy and the management of the territory (Berkes and Folke, 1998; Hahn et al., 2006), as is a requirement in the governability approach also. Accordingly, Chilean planners may decide to consider the actions proposed within the context of the adaptive governance literature to reach governability.

First, there is the approach of including local ecological knowledge in formal management directed by the state (Evans et al., 2011). In this case, governmental institutions become mediators between scientists, managers and the community. The institutions mediate the integration of knowledge based on research in the management of the territory; they mediate the integration of local knowledge systems into the decision-making process; and intercede to ensure the early integration of local knowledge and processes in planning. Second, the option is changing the decision-making process in an emergency scenario (Brunner et al., 2005). The proposal in this case is to avoid a single or hierarchical authority that decides on the relevant issues. This proposal is born from the idea that in governments with a fragmented political-administrative system and without interrelation among planning and emergency instruments—as is the case of the regulatory framework for the Chilean coast—local initiatives from educated and informed community groups can help to overcome the gridlock and bureaucracy in the management of the territory. Accordingly, it will be possible to be able to integrate and balance the different interests of the community.

Another important finding of this study, that can be considered a challenge if the aim is to reach governability, is that the regulatory framework is not considering technical data during the decision making process. The consideration of the scientific knowledge has been discussed as particularly important to ensure resiliency in developing contexts (Brunner et al., 2005). Our results revealed that the regulatory framework does not set out binding regulations between emergency and planning instruments, and this is particularly problematic for the definition of land use and risk zones. We found that most of the regulations with a negative orientation to resilience were related to the lack of consideration of the tsunami inundation line. Much built and unbuilt infrastructure proposed to cope with the disaster is proposed under this line. The way in which this information and that from emergency planning instruments is included in the land use definition of the territory should be re-evaluated because in the process, science is being left aside. Even though there is an extensive amount of research in Chile developed with the aim of modeling tsunamis along the Chilean coast (e.g., Aranguiz et al., 2016), results of this type of research seems to have not been integrated in planning as our results suggest. In coastal areas the achievement of this includes innovation in policy, by integrating and balancing multiple interests, and in science, with a more integrative and interpretable science (Brunner et al., 2005). This last aspect sets out a proposal for scientists in order to consider the subjective viewpoints or experiences of the community to tackle the uncertainties that evolve from applied science. The findings of our study support the proposal of reviewing how science is being integrated in planning in a disaster scenario, because as it is, has not been useful to focus the regulations into a resilience and governability path.

5 Conclusion

Overall, our study revealed a trend towards a governance over a governability approach in the case studies along the Chilean coast. This approach includes little flexibility and redundancy in the manner in which the territory is administered in case of a disaster, regardless of the fact that the regulations are establishing norms with a positive over a negative effect in the multiple resilience dimensions of the territory. In particular, the governance approach implies here that the resilience capacity of the territory is highly dependent on the capabilities and decisions taken by the emergency and planning institutions, while the community capacity and expertise to adapt in case of disaster receives lesser attention. This is an important finding to consider in future coastal planning in developing countries. The changes generated by natural hazards and a diversity of stakeholders with different residential, commercial and industrial needs and aims are resulting in a growing and sustained pressure on coastal zones, reaching todays dramatic proportions. Our findings indicate that responding to this phenomenon not only implies the action of planners, but also that of police makers, scientists, emergency managers and the community. All should get involved into an urgent revision of the codes that sustain the coastal occupation and its resiliency without pre-established hierarchies or inequalities in the interference that each one has. Then, if the aim is to reach a governability approach that ensures resiliency, the challenge for local governments in coastal environments is to reconcile multiple interests, with multiple resilience dimensions, in a regulatory framework that assures complementation and equilibrium among all.

References

Aguilar, L.F., 2013. Tercera parte. Gobernanza. In: Aguilar, L.F. (Ed.), Gobierno y Administración Pública. Fondo de Cultura Económica, Mexico, pp. 271–376.

Andreu, J., 2000. Las Técnicas de Análisis de Contenido: Una Versión Actualizada. vol. 10 Fundación Centro Estudios Andaluces, Universidad de Granada, España.

Aranguiz, R., Gonzalez, G., Gonzalez, J., Catalan, P., Cienfuegos, R., Yagi, Y., et al., 2016. The 16 September 2015 Chile tsunami from the post-tsunami survey and numerical modeling perspectives. Pure Appl. Geophys. 173 (2), 333–348.

Berkes, F., Folke, C., 1998. Linking Social and Ecological Systems, Management Practices and Social Mechanisms for Building Resilience. Cambridge University Press, Cambridge, UK.

Blanco, H., Alberti, M., Olshansky, R., Chang, S., Wheeler, S.M., Randolph, J., et al., 2009. Shaken, shrinking, hot, impoverished and informal: emerging research agendas in planning. Prog. Plan. 72, 195–250.

Botero, C., Milanés, C., Inciarte, L., Arrizabalaga, M., Vivas, O., 2015. Aportes para la gobernanza marino-costera. Gestión del riesgo, gobernabilidad y distritos costeros. Universidad Sergio Arboleda, Bogotá.

Brunner, R., Steelman, T.A., Coe-Juell, L., Cromley, C.M., Edwards, C.M., Tucker, D.W., 2005. Adaptive Governance: Integrating Science, Policy, and Decision Making. Columbia University Press.

Camou, A., 2011. Gobernabilidad y Democracia. vol. 6 Instituto Federal electoral, México.

Cutter, S.L., Ash, K.D., Emrich, C.T., 2014. The geographies of community disaster resilience. Glob. Environ. Chang. 29, 65–77.

Demuth, J.L., Morss, R.E., Lazo, J.K., Trumbo, C., 2016. The effects of past hurricane experiences on evacuation intentions through risk perception and efficacy beliefs: a mediation analysis. Weather Clim. Soc. 8 (4), 327–344.

Djalante, R., 2012. Adaptive governance and resilience: the role of multi-stakeholder platforms in disaster risk reduction. Nat. Hazards Earth Syst. Sci. 12, 2923–2942.

Evans, L., Brown, K., Allison, E., 2011. Factors influencing adaptive marine governance in a developing country context: a case study of southern Kenya. Ecol. Soc. 16 (2).

Forbes, K., Broadhead, J., 2007. The Role of Coastal Forests in the Mitigation of Tsunami Impacts. FAO, Bangkok.

Gall, M., Cutter, S.L., Nguyen, K., 2014. Governance in Disaster Risk Management (Vol. IRDR AIRDR Publication No. 3). Integrated Research on Disaster Risk, Beijing.

Hahn, T., Olsson, P., Folke, C., Johansson, K., 2006. Trust-building, knowledge generation and organisational innovations: the role of a bridging organisation for adaptive co-management of a wetland landscape around Kristianstad, Sweden. Hum. Ecol. 34 (4), 573–592.

Herrmann, G., 2015. Urban planning and tsunami impact mitigation in Chile after February 27, 2010. Nat. Hazards 79 (3), 1591–1620.

Holling, C.S., Gunderson, L.H., Ludwig, D., 2001. In quest of a theory of adaptive change. In: Panarchy: Understanding Transformation in Human and Natural Systems. Island Press, Washington, pp. 3–23.

Kooiman, J., 2005. Gobernar en gobernanza. In: Cerrillo, I.M. (Ed.), La gobernanza hoy: 10 textos de referencia. Instituto Nacional de Administración Pública, Madrid, pp. 57–83.

Krippendorff, K., 2004. Content Analysis: And Introduction to its Methodology, second ed. Sage Publications Inc., California.

Murray-Tuite, P., Yin, W., Ukkusuri, S.V., Gladwin, H., 2012. Changes in evacuation decisions between Hurricanes Ivan and Katrina. Transp. Res. Rec. 2312, 98–107.

Norris, F., Stevens, S., Pfefferbaum, B., Wyche, K., Pfefferbaum, R., 2008. Community resilience as a metaphor, theory, set of capacities and strategy for disaster readiness. Community Psychol. 41, 127–150.

Stein, R.M., Dueñas-Osorio, L., Subramanian, D., 2010. Who evacuates when hurricanes approach? The role of risk, information, and location. Soc. Sci. Q. 91 (3), 816–834.

Tierney, K., 2012. Disaster governance: social, political, and economic dimensions. Annu. Rev. Environ. Resour. 37, 341–363.

Tomassini, L., 1996. Gobernabilidad y políticas públicas en América Latina. Banco Interamericano de Desarrollo, Washington, DC.

Trumbo, C., Meyer, M.A., Marlatt, H., Peek, L., Morrissey, B., 2013. An assessment of change in risk perception and optimistic bias for hurricanes among Gulf Coast residents. Risk Anal. 34 (6), 1013–1024.

Vale, L.J., Campanella, T.J. (Eds.), 2005. The Resilient City, How Modern Cities Recover from Disaster. Oxford University Press, New York.

Villagra, P., Quintana, C., 2017. Disaster governance for community resilience in coastal towns: Chilean case studies. Int. J. Environ. Res. Public Health 14 (9). pii: E1063.

Villagra, P., Herrmann, M.G., Quintana, C., Sepúlveda, R.D., 2017. Community resilience to tsunamis along the Southeastern Pacific: a multivariate approach incorporating physical, environmental, and social indicators. Nat. Hazards 88 (2), 1087–1111.

Walker, B., Salt, D., 2006. Resilience Thinking: Sustaining Ecosystems and People in a Changing World. Island Press, Washington.

Walker, B., Holling, C.S., Carpenter, S.R., Kinzig, A., 2004. Resilience, adaptability and transformability in social–ecological systems. Ecology and Society 9 (2). ART. 5.

Wu, H.C., Lindell, M.K., Prater, C.S., 2012. Logistics of hurricane evacuation in Hurricanes Katrina and Rita. Transport. Res. F: Traffic Psychol. Behav. 15 (4), 445–461.

Chapter 9

Harmonizing policies to enhance cross-border regional resilience of the Guangdong-Hong Kong-Macau Greater Bay Area

Qingqing Feng, S. Thomas Ng, and Frank J. Xu
Department of Civil Engineering, The University of Hong Kong, Hong Kong, China

1 Introduction

Improving community and infrastructure resilience has become the main focus of many metropolis and regions to pursue sustainable development and tackle multi-hazards including natural disasters, climate change, man-made shocks, and infrastructure and environmental challenges. To address those issues, the international society has made concerted efforts by developing the post-2015 development agenda which is founded on the basis of the Paris Agreement (UNFCCC, 2015), Sendai Framework, Sustainable Development Goals (SDGs) and New Urban Agenda (Roberts et al., 2015). The Paris Agreement strives to strengthen the global response to threats posed by climate change while the Sendai Framework sets out seven clear targets and four priorities from the prospective of disaster risks reduction. The SDGs represent a coherent way of thinking about sustainable development from the entwined economic, social and environmental perspectives. The New Urban Agenda presents a critical framework to mobilize cities and regions to promote resilience and environmental sustainability in urban settings.

While goals and targets for adaptation and resilience have been set out at the international level, governments in the Guangdong-Hong Kong-Macau Greater Bay Area, China (hereinafter referred to as the Greater Bay Area), which comprises nine cities of Guangdong Province,[a] the Hong Kong Special Administrative Region (SAR) and the Macao SAR, are endeavoring to formulate and implement local policy instruments to correspond with their unique governance characteristics and to fulfill their sustainability goals and targets. For example, the Shenzhen government has developed the "*Shenzhen City Sustainable Development Plan (2017-2030)*" on the basis of the SDGs; and Hong Kong has also compiled its future development strategic plan "Hong Kong 2030+" to align with the UN's New Urban Agenda.

The concept of "the Greater Bay Area" in China was formally put forward in 2015 in developing the country's "One Belt and One Road" initiative, which is a crucial part and motive force of China's on-going regional development strategy. Therefore, it is of paramount importance to build a liveable world-class city cluster with due consideration of regional resilience. Cities within the Greater Bay Area are not only geographical neighbors but they also culturally connected with a long history. Currently, the governments within the area are striving to realize better interconnectivity of infrastructure network, and promote joint development and comprehensive cooperation to enhance regional resilience. However, Hong Kong, Macao and the other nine cities in the area have adopted different political systems; not to mention that the governance structures, legal systems, political systems and currencies also vary. Public policies, as a course of actions being adopted by authorities to address problems or realize a goal, are significant *en route* to regional resilience (Pal, 2005). Thus, it is a great challenge for governments in the Area to formulate and implement synergised policies to improve the cross-border regional resilience due to the complexity and heterogeneity of the three places.

This chapter aims to provide a theoretical basis for policy makers of the Greater Bay Area to formulate a harmonized blueprint for planning and development of a resilient city cluster. Following an introduction to the relevant research studies on regional resilience, the widely recognized "NATO" scheme (Nodality, Authority, Treasure and Organization) and the enhanced framework for adaptation and resilience policy analysis (FARP) developed by the authors, as well as the common

[a] The nine cities of Guangdong Province include Guangzhou, Foshan, Zhaoqing, Shenzhen, Dongguan, Huizhou, Zhuhai, Zhongshan and Jiangmen.

Strengthening Disaster Risk Governance to Manage Disaster Risk. https://doi.org/10.1016/B978-0-12-818750-0.00009-X

hazards of Hong Kong, Macao and the other nine cities in Guangdong Province are listed. Then, the policy instruments of the three regions in tackling these common hazards are identified and further grouped into different matrixes with community capitals and the NATO scheme being the horizontal and vertical dimensions. Based on the matrices, the extant policy instruments are qualitatively appraised and compared to derive insightful analysis and suggestions.

2 Related works

2.1 Regional resilience

Resilience is defined as the ability of an entity "*to recover from setbacks, adapt well to change, and keep going in the face of adversity*" (Ovans, 2015). Due to the growing frequency and severity of natural hazards, man-made hazards, and climate extremes that threaten world's development and human's well-being (Rockefeller Foundation and Arup, 2014; Takagi et al., 2018), various local and global initiatives have been launched to improve community's ability to "*plan for, absorb, respond to, recover from, and more successfully adapt to new adverse events*" (NRC, 2015). Despite great research efforts being made to community and city resilience, there are limited studies focusing on regional resilience within the city clusters, especially those cross-border city clusters like the Greater Bay Area and the Yangtze River Delta regions in China, the Singapore-Johor-Riau (SIJORI) Growth Triangle, etc. Unfortunately, regions have become more permeable to the effects of what were once thought to be external processes which necessitates the application of the concept of "regional resilience" (Christopherson et al., 2010). As a consequence, different cities within a city cluster will need to coordinate and cooperate to build up resilience capability against all forms of common hazards they may encounter.

As a core part of China's national strategic urban planning, the resilience of the Greater Bay Area should be given due consideration for its substantial prosperity and development for the years to come. The National New-type Urbanisation Plan also emphasizes the need to optimize and enhance the Pearl River Delta (PRD) region to make it a global urban agglomeration (Ng et al., 2018). However, it is understood that building regional resilience of city clusters is indeed more challenging than uplifting the resilience of a single city, especially for the cross-border regions with heterogeneous political and legal systems.

2.2 The FARP framework

Framework for adaptation and resilience policy analysis (FARP) is a framework developed by the authors to assist governments in conducting policy analysis and to help them develop new policy instruments for improved adaptation and resilience capacities as well as to pave ways for the implementation of post-2015 development agenda (Ng and Xu, 2016). The enhanced FARP framework is shown in Fig. 9.1. The following sections not only compare the policies, strategies and

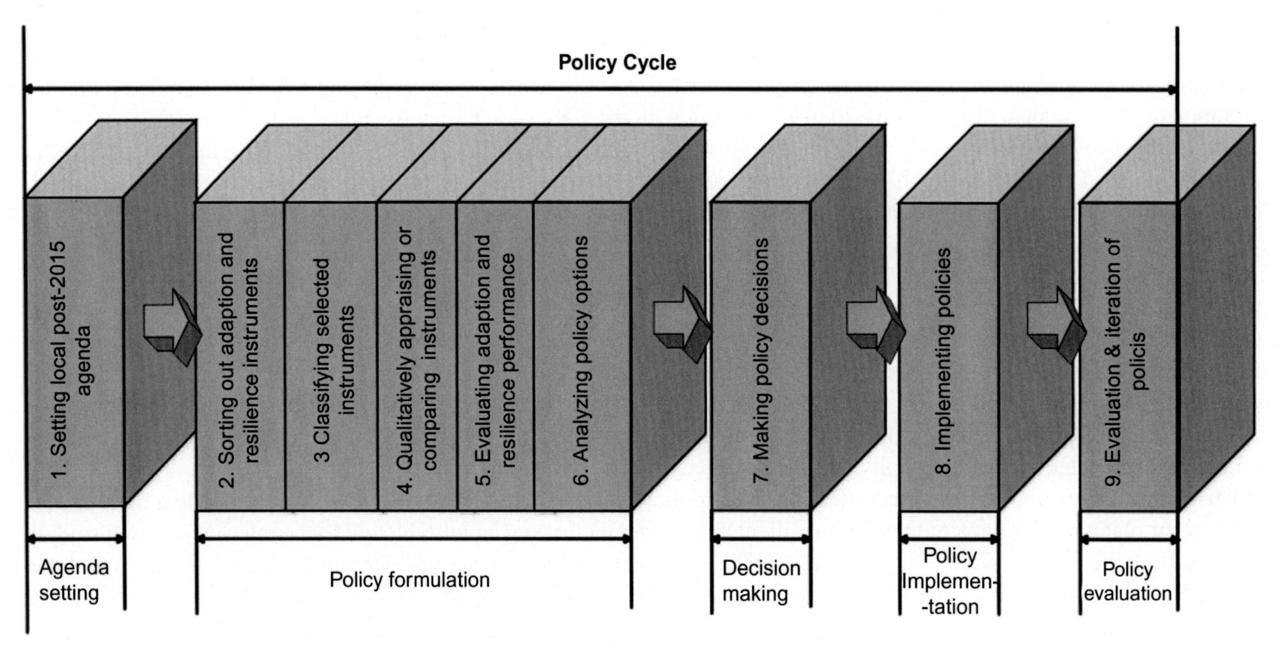

FIG. 9.1 Schematic diagram of the FARP framework.

plans of the three areas in the Greater Bay Area, but also investigate the gaps and solutions to align their urban and regional resilience improvement practices.

2.3 Policy instruments and NATO scheme

The NATO scheme, on which the FARP framework is based, was first developed by Hood (1986) who contemplated that governments essentially have four types of tools and resources, namely: nodality, authority, treasure and organization (NATO), to either affect or inhibit changes in social environment. The specific descriptions of each type of resources are shown as follows (Henstra, 2016).

- Nodality: Use knowledge generation, sharing and mobilization as well as information dissemination, education and training to achieve the policy target.
- Authority: Use legitimate power to permit, prohibit or command the target populations.
- Treasure: Use monetary resources and financial incentives, e.g., direct program spending, subsidy, fund, tax, or credit to induce targets to adopt to the policy-preferred behavior.
- Organization: Governments implement policy objectives directly by delivering the services or by incorporating policy principles into government operations.

In the last three decades, NATO has received particular attention and a number of variations of NATO have been developed to understand policy-making process, as it offers an attractive way and a great potential to break down complex public policies into constituent components and thereby provides a more prosaic metaphor to solve intricate policy problems (Margetts and Hood, 2016). The FARP framework could, therefore, be utilized to compare, analyze, make, and implement cross-border regional resilience policies of the Greater Bay Area in China.

3 Data and methodology

Using the FARP framework as a resilience policy-making toolkit, extant policy instruments related to regional resilience that are being used or having been implemented by the Greater Bay Area member governments are identified, classified and qualitatively appraised.

First, the common hazards facing member cities of the Greater Bay Area are determined. The top eight common challenges/hazards of the region are coastal flooding, storm surge, typhoon, rainfall flooding, sea-level rise, climate change, poor air quality and environmental degradation. These hazards are recognized through two rounds of interview with different policy-making stakeholders by using a list of challenges adopted by the 100 Resilient Cities initiative. The first round was undertaken in August 2017 via crowdsourcing. Over 500 effective online questionnaires were collected from local citizens in Greater Bay Area. The questionnaires can be answered by clicking website links or scanning QR code. Based on the result of the questionnaires, the second interview round was undertaken in November 2017 with selected interviewees. Of the 16 interviewees involved, six are from government departments, four are from public organizations or non-government service providers, two are from consultancy companies and the remaining four are academic members. The basic profile of these interviewees is shown in Table 9.1 below.

Second, the policy instruments promulgated by the Greater Bay Area cities to deal with the aforementioned hazards are identified from relevant government websites or acquired directly from various responsible institutions. These policy instruments are developed, enforced and monitored by different governmental institutions affiliated to different cities (see Fig. 9.2).

Third, the identified policy instruments are broken down into different constituent components and grouped into different matrixes/maps, with the community capitals and NATO scheme serving as the horizontal and vertical dimensions. Community resources or assets that are able to be invested to create new ones can be upgraded to become various "capitals." Some scholars found it extremely useful to divide community capitals into seven types when conducting holistic analysis of a community, including natural, cultural, human, social, political, financial and built capitals (Flora, 2018). The resilience policy matrixes obtained for Hong Kong, Macau and the Guangdong Province are shown in Tables 9.2, 9.3 and 9.4, respectively.

Finally, qualitative analysis and comparison of those resilience policies found in these three places were carried out. Policy interdependencies and gaps can be clearly identified and presented to different stakeholders. On the other hand, different policy-makers can also leverage the matrices to align with their policy making and execution processes and to evaluate the performance of any resilience policy.

TABLE 9.1 Particulars of interviewees.

Code	Sector	Institution	Position	Experience (work year)
1	Water-related	Government department	Assistant director	>10
2	Water-related	Government department	Senior engineer	>5
3	Drainage service	Government department	Senior engineer	>5
4	Drainage service	Government department	Senior O&M engineer	>5
5	Environment-related	Government department	Senior engineer	>5
6	Emergency management	Government department	Senior engineer	>5
7	Electricity	Service provider	Senior operation engineer	>5
8	Electricity	Service provider	Senior operation engineer	>5
9	Railway	Service provider	Senior operation engineer	>5
10	Railway	Service provider	Senior operation engineer	>5
11	Utilities	Consultancy	Senior engineer	>10
12	Utilities	Consultancy	Engineer	>3
13	Civil Engineering	University	Associate Professor	>15
14	Civil Engineering	University	Professor	>20
15	Public Policy	University	Post-doctor	>5
16	Public Policy	University	Post-doctor	>5

	Hong Kong	Macao	Guangdong
Natural Disasters	Drainage Service Department; Civil Engineering and Development Department Security Bureau; Hong Kong Observatory	Macao Meteorological and Geophysical Bureau (SMG); Land, Public works and Transport Bureau; Marine and Water Bureau(DSAMA) Secretariat for Security	Emergency Management Office of Guangdong
Environment Problems	Evironment Bureau; Environmental Protection Department	Evironmental Protection Bureau (DSPA)	Guangdong provincial department of environmental protection; Environmental Protection Bureau of each city
Climate change	Transport Department; **the Inter-department Working Group on Climate Change;** Building Department; Environment Bureau; Environmental Protection Department	Evironmental Protection Bureau (DSPA); **Inter-department Task Force on Climate Change**	Guangdong provincial department of environmental protection; Mistry of Housing and Urban-Rural Development; Energy Bureau; Environmental Protection Bureau of each city

FIG. 9.2 Government institutions affiliated to different cities.

The FARP framework can be applied iteratively for establishing and optimizing community and regional resilience policies. All the steps can be considered as in a small-scale cycle embedded in a bigger-scale policy cycle. This research analyses and compares the extant resilience polices of the Greater Bay Area cities' by applying the "policy formulation" link of the FARP framework. The preliminary findings and observations are presented in the next section.

TABLE 9.2 Classification of policy instruments in Hong Kong.

Capital	Nodality	Authority	Treasure	Organization
Natural	A Clean Air Plan for Hong Kong; the PRD Regional Air Quality Management Plan; Low Carbon Living Calculator; Hong Kong's Climate Action Plan 2030+; Energy Saving Plan for Hong Kong's Built Environment 2015–2015+; Hong Kong's Climate Change Report; Deepening Energy Saving In Existing Buildings	Air Pollution Control Ordinance; Water Pollution Control Ordinance; Flood Protection Standards; Environmental Impact Assessment Ordinance	Build Sludge Treatment Facility T-Park to recover energy; Polluter Pays Principle; Sustainable Development Fund; Pilot Green Transport Fund; Environment and Conservation Fund	The Hong Kong-Guangdong Joint Working Group on Sustainable Development & Environmental Protection; The 2016–2020 co-operation agreement on environmental protection; The Harbour Area Treatment Scheme
Human			HKSAR Emergency Response System	
Social/political/financial	N/A			
Built	DSD Outreach Educational Programme; Education Kit for Liberal Studies on Flood Prevention and Sewerage; Drainage Master Plan; Sewerage master plans; Public Transport Strategy Study	Land Drainage Ordinance; Building (Energy Efficiency) Regulation; Waste Disposal Ordinance; Buildings Energy Efficiency Ordinance	Flood warning system; Contingency Plan for Natural Disasters; Emergency and Storm Damage Organization; Sustainable Development Fund; Pilot Green Transport Fund; The Buildings Energy Efficiency Funding Schemes; Energy Efficiency Registration Scheme for Buildings	Preventive maintenance of the storm water drainage system; Landslip Prevention and Mitigation Programme

4 Preliminary findings and observations

The preliminary results are presented as follows:

- Hong Kong, Macao and Guangdong Province all recognize the importance of drawing up long-term roadmaps for tackling global climate change and mitigating natural hazard risks and environmental degradation.
- The heterogeneity of government structure within the Greater Bay Area poses a great challenge to aligning and orchestrating various departments and bureaux.
- There are few cooperative initiatives in the Greater Bay Area to enhance regional resilience. For example, the Pearl River Delta (PRD) regional air monitoring network has been established to tackle the air pollution problem, and the Regional Air Quality Management Plan has also been drawn up to steer and coordinate the actions of environmental-related bureaux and departments within the Greater Bay Area region.
- A policy goal cannot be achieved by a single department or city due to the growing complexity of common hazards, the interdependency of cities' infrastructure, and the mobility of neighboring citizens. For example, tackling climate change hazard not only requires individual efforts but also the coordination and cooperation among the entwined government institutions. The establishment of inter-departmental working groups in Hong Kong and Macao sets a good example to enhance regional resilience.
- The Greater Bay Area needs a set of aligned policy instruments to enhance cross-border regional resilience. Policy instruments belonging to different categories of "NATO" scheme or different community capitals can be utilized collectively to drive behavioral changes of citizens for improved resilience capacity.

TABLE 9.3 Classification of policy instruments in Macao.

Capital	Nodality	Authority	Treasure	Organization
Natural	Macao Environmental Protection Plan (2010–2020); Air quality monitoring and forecast; Guangdong, Hong Kong and Macao pearl river delta regional air monitoring network; Launch action plans to promote green travel; Carbon footprint calculator; Toward a thematic strategy for the climate change in Macao Special Administration Region	Standards for unleaded gasoline and light diesel for vehicles	Implementing subsidy schemes to phase out two-stroke motor vehicles; All kinds of severe weather warnings including rainstorm, thunderstorm, tropical cyclone, strong monsoon, storm surge, etc.; Environmental Protection and Energy Saving Fund	Guangdong, HK and Macao meteorological technology seminar and meteorological cooperation conference; Implementing e-government to reduce carbon emissions; Public sector environmental procurement guidelines; Inter-departmental Task Force on Climate Change
Human			Evacuation plan for low-lying areas with storm surge during typhoon; Ten Year Plan for Disaster Prevention and Mitigation in Macao (2019–2028)	Building Emergency Command Application Platform
Social/political/financial	N/A			
Built	Public sector green travel guidelines; Macao Land Transport Policy (2010–2020)	Standards for unleaded gasoline and light diesel for vehicles; Mandatory inspection of light bus and motorcycle; Regulations on exhaust emission standards for new automobiles imported	Implementing subsidy schemes to phase out two-stroke motor vehicles; Tax incentives for new motor vehicles that meet environmental standards; Macao Green Hotel Award; Promoting Citizens' Recycling by giving in return supermarket coupon; SME funding scheme for installing flood gates/anti-flooding lifting platform	Construction of Tidal Barriers and Integrated Water Control Measures for the Inner Harbour in Macao

- Most adaptation and resilience practices are driven or led by the governments with their major focus being on the built and natural capitals. Treating all capitals as a whole, engaging citizens and deploying balanced "Nodality," "Authority," "Treasure," "Organization" policy instruments would help improve regional resilience.
- The policy philosophy of three places in formulating and executing resilience policy instruments varies from each other. In Hong Kong and Macao, local policy agendas, such as the "Hong Kong 2030+," are often developed by one responsible department first, and then referred by other departments and bureaux to implement policy instruments and practices. In Guangdong Province, the provincial government sets the policy agendas according to the national strategies and plans, and its nine cities then formulate their local policy instruments and implementation plans according to their individual needs and characteristics. With such a complex policy system, it is difficult to list out all the policy instruments, and even more challenges for combination of their constituent components.
- The resilience policy matrixes and the FARP framework hold promising potentials to make the analysis and comparisons of adaptation and resilience policies simple and easy-to-understand. Nevertheless, automated resilience policy analysis tools shall be developed to evaluate specific policies with reference to the FARP framework. Iterative policy evaluations and performance appraising will help facilitate the next cycle of resilience policy development.

TABLE 9.4 Classification of policy instruments in Guangdong Province.

Capital	Nodality	Authority	Treasure	Organization
Natural	The PRD Regional Air Quality Management Plan; Guangdong Province Air Pollution Prevention Action Plan (2014–2017); Guangdong province's climate change plan	Regulations on the prevention and control of air pollution in Guangdong province; By 2020, the water area of cities in the Pearl River Delta and coastal areas will be no less than 10%	Implementation of contract energy management tax incentives; Establish systems such as green credit, green securities, and environmental pollution liability insurance, and incorporate corporate environmental information into the bank's credit information system	The Hong Kong-Guangdong Joint Working Group on Sustainable Development & Environmental Protection; The 2016–2020 co-operation agreement on environmental protection; Consultation and cooperation mechanism on joint control of air pollution in PRD
Human			Guangdong Public Emergency Warning Information Release System	Flood prevention and drought prevention emergency plan for Guangdong Province
Social/political/financial	N/A			
Built	Developing green traffic	All coal-fired units in the PRD implement the "Emission Standard for Air Pollutants from Thermal Power Plants"; Deepen the governance of industrial sources and promote desulfurization and denitrification; Strictly control the use of petroleum coke; Outlaw "small, decentralized, messy, polluted" enterprises	Increase investment in the construction and renovation of urban drainage and flood prevention facilities; Developing green traffic	Improve the emergency mechanism

- The cooperation among the cities of the Greater Bay Area to improve regional resilience is still at its infancy stage. Besides aligning the resilience policies, involved cities also need to strengthen their collaboration in establishing the resilience management institutions, sharing resilience knowledge and information, developing resilience evaluation standards and supporting systems, engaging citizens, and building inclusive and resilient communities.

5 Conclusions

To build resilience into the Greater Bay Area (Hong Kong, Macao and Guangdong) in China requires the development of tailor-made community resilience policies and a coherent framework for cross-border cooperation. This paper extends the FARP framework developed by the authors and applies it to identify, classify, analyze and compare the policy instruments of the Greater Bay Area pertinent to traditional disaster and emergency management, climate change and sustainable development. Using the policy matrixes with the "NATO" (Nodality, Authority, Treasure and Organization) scheme as the horizontal dimensions and the community capitals as the vertical dimensions, stakeholders can easily understand the gaps and strengths of the existing resilience policies and practices. The extant plans, policies and cooperation initiatives of the three places in tackling the eight common hazards were identified and examined. The preliminary findings showed that more efforts are needed to strengthen the cross-city cohesiveness of policy-making. Besides, inter-departmental cooperation is imperative for improving cross-border regional resilience, in particular to ensure the consistency and completeness of relevant polices and to derive harmonized plans for evaluating and optimizing the policies.

To effectively analyze, compare and appraise the resilience policies of the Greater Bay Area member governments and maximize the potentials of applying the FARP framework to resilience policy analysis, future research should be carried out, and these include: (i) clarifying the comprehensive resilience policy scenarios to meet the requirements of different stakeholders; (ii) exploring the barriers to and possibilities of setting up a distinctive organization to orchestrate the development and execution of regional resilience policies and the operating mechanisms of the organization; (iii) developing a supporting system to automate policy analysis procedures and to share resilience improvement knowledge and practices; (iv) conducting detailed case studies; and (v) integrating other contemporary approaches to the public policy.

Acknowledgments

The authors would like to thank the Research Grants Council of the HKSAR Government for funding this work under the General Research Fund (Project Nos.: 17202215, 17248216 & 17204017).

References

Christopherson, S., Michie, J., Tyler, P., 2010. Regional resilience: theoretical and empirical perspectives. Camb. J. Reg. Econ. Soc. 3 (1), 3–10.

Flora, C.B., 2018. Rural communities: Legacy + change. Routledge, London.

Henstra, D., 2016. The tools of climate adaptation policy: analysing instruments and instrument selection. Clim. Pol. 16 (4), 496–521.

Hood, C., 1986. The Tools of Government. Chatham House, Chatham, NJ.

Margetts, H., Hood, C., 2016. Tools approaches. In: Peters, B.G., Zittoun, P. (Eds.), Contemporary Approaches to Public Policy. Theories, Controversies and Perspectives. Palgrave Macmillan, Basingstoke, pp. 133–155.

Ng, T.S.T., Xu, J., 2016. Comparing and evaluating Hong Kong's adaptation and resilience policies. In: Proceedings of the 6th International Conference on Building Resilience, New Zealand, 7–9 September.

Ng, S.T., Xu, F.J., Yang, Y., Lu, M., Li, J., 2018. Necessities and challenges to strengthen the regional infrastructure resilience within city clusters. Procedia Eng. 212, 198–205.

NRC, 2015. Developing a Framework for Measuring Community Resilience. National Research Council, The National Academies Press, Washington, DC.

Ovans, A., January 2015. What resilience means, and why it matters. Harv. Bus. Rev.

Pal, L.A., 2005. Beyond Policy Analysis: Public Issue Management in Turbulent Times. Thomson Nelson, Scotland.

Roberts, E., Andrei, S., Huq, S., Flint, L., 2015. Resilience synergies in the post-2015 development agenda. Nat. Clim. Chang. 5, 1024–1025. https://doi.org/10.1038/nclimate2776.

Rockefeller Foundation & Arup, 2014. City Resilience Framework. ARUP & The Rockefeller Foundation, London.

Takagi, H., Xiong, Y., Furukawa, F., 2018. Track analysis and storm surge investigation of 2017 Typhoon Hato: were the warning signals issued in Macau and Hong Kong timed appropriately? In: Georisk: Assessment and Management of Risk for Engineered Systems and Geohazards, pp. 1–11.

UNFCCC, 2015. Adoption of the Paris Agreement. United Nations Framework Convention on Climate Change, Washington, DC.

Chapter 10

Participatory Geographic Information Systems for integrated risk analysis: A case of Arequipa, Peru

Carlos Zeballos-Velarde
Universidad Catolica de Santa Maria, Arequipa, Peru

1 Introduction

In the coming years, it is likely that climate change will pose an increased number of climate-related risks, affecting the most vulnerable settlements, causing destructive floods, failures in the supply of water and energy and the subsequent economic impact on these communities (Muller, 2007). In most developing countries in particular, large segments of vulnerable populations are located in hazard prone areas, and the occurrence of disaster events "can generate greater impacts on populations and territories that are frequently characterized by extremely vulnerable conditions" (Giovene di Girasole and Cannatella, 2017).

In Peru, the National Centre for Estimation, Prevention and Reduction of Disaster Risks (CENEPRED), proposes to identify the hazards by evaluating the intensity, magnitude, recurrence period and the level of susceptibility of natural phenomena, and conduct the analysis of components that affect vulnerability which is determined by exposure, fragility and resilience, and the identification of potentially vulnerable elements. The level of risk faced by an area is usually calculated by the sum of hazard and vulnerability analysis (CENEPRED, 2014). In order to carry out an analysis and evaluation of vulnerabilities of a locality, it is necessary to identify its conditions of exposure, fragility and resilience, which will determine the multiple vulnerability models that allow for identification of critical sectors within a community (Ministry of Environment, Peru, 2015). Therefore, it is important to highlight that there are several types of vulnerability that converge in the determination of these critical elements: physical, social, economic and environmental vulnerability (Grupo GEA, 2008; UNISDR, 2009).

Given the need to identify the areas that are most prone to disasters, the use of Geographic Information Systems has proven to be useful for the development and mapping of hazards, vulnerabilities and risks, basing their success on the accuracy and relevance of the data that feed these analyses. However, the cost of surveying large areas and processing a big amount of information corresponding to all jurisdictions is quite high, which can be unaffordable, particularly in developing countries (Giovene di Girasole and Cannatella, 2017).

In Peru, most municipalities do not have the capacity for risk analysis; and hazard mapping is often carried out by government agencies, on a very broad and not very detailed scale. In addition, the use of GIS in these entities has a cartographic rather than an analytical nature. Therefore, Participatory GIS (PGIS) is suggested to serve as an alternative that can incorporate traditionally marginalized sectors in the use of GIS tools (Radil and Jiao, 2016).

In Peru, PGIS can become an opportunity to encourage the involvement of the population in the discussion and identification of the most hazardous areas in their community. In addition, it can serve as a reliable source for urban planners to fill in the information gap as well as providing the capacity, which the existing Peruvian risk modelers and GIS specialists lack, in terms of hazard risk analysis and mapping. In addition, PGIS can promote dialogue between the community, urban planners and decision-makers, which will be useful in future stages of disaster risk management.

In this context, this research aims to propose a methodology that emphasizes citizen participation by the use of PGIS, as a fundamental input method to assess disaster risk management. The Jacobo Hunter district was chosen as a pilot area, since it is representative of the hazards that are found in other areas of the city.

Strengthening Disaster Risk Governance to Manage Disaster Risk. https://doi.org/10.1016/B978-0-12-818750-0.00010-6

2 Participatory Geographic Information Systems

Participatory mapping is a concept referred to the different ways in which social groups capture a particular vision of the world through maps, in order to communicate ideas, knowledge, experiences, and ideals (Tane and Wang, 2007). The International Fund for Agricultural Development defines it as "the creation of maps by local communities, often with the involvement of supporting organizations including governments, non-governmental organizations, universities, and other actors engaged in the development and land-related planning" (IFAD, 2009).

Several authors have underlined the concept of participatory mapping as a transformative and creative process which can be useful in several activities such as "delineating territorial boundaries, identifying important places that sustain livelihoods and quality of life, and communicating preferences about future land use" (Browna and Kyttä, 2018). In addition, the need for more participation in the making of maps by the people who are affected by them is emphasized (McCall, 2014). However, others have expressed concern about what they consider "abuse of participation," which can turn out to be time-consuming, create unfulfilled expectations, become ineffective and even produce ethical problems if the information is misused (Chambers, 2006).

Nevertheless, there is an important amount of evidence of the benefits of the combination of participatory mapping and the powerful characteristics of geographic information, which is called Participatory GIS or PGIS. Tane and Wang suggested that PGIS "combines both quantitative and qualitative methods and relates them through community development processes and regional initiatives to unite conventional knowledge with community cultural intelligence. [...] This allows participatory surveys and data collection within local communities in ways that reflect actual and perceived situations in the villages" (2007). Accordingly, McCall (2014) proposes five key principles of PGIS, which are (a) empowering through participation; (b) aiming for characteristics of flexibility and adaptiveness, and ongoingness; (c) respect for local knowledge; (d) awareness and utilization of local and locational specificity; and, (e) being reflexive and self-aware in its recognition of agency and power.

However, there are different approaches regarding PGIS, depending on whether the stress is put in the participation or the use of Geo-informatics. For example, in the Northern countries, the use of Public Participation GIS (PPGIS), whose main goal is to use GIS "to provide information that can strengthen involvement of communities or marginalized groups in decision-making" (Chingombe et al., 2015), while in the Global South, "PGIS practice has emerged as an intersection of participatory progressive development and GIT&S through the integration of low and high tech spatial information management applications" (Rambaldi et al., 2006) and it is highly dependent on external technology inputs.

In terms of disaster risk reduction, PGIS seeks to develop risk mapping, modeling and monitoring in an integrated vision that involves the actors within the various phases of risk management: the proactive, reactive and corrective phase. PGIS tends to focus on "user-friendly methods of collecting, storing, retrieving, transforming and displaying spatial data for wide ranging purposes like locating positions in the field, sustainable development planning, integrated resource management, monitoring assets, assessing natural hazards, as well as many other applications" (Tane and Wang, 2007).

3 The site

Arequipa, the second city of Peru, with a population of 800,000 inhabitants, has experienced in the last decades a chaotic and unplanned growth, carried out by migrants or people of low socioeconomic level, often occupying high risk areas. As a consequence, human and material losses due to the occurrence of disasters such as floods, landslides and earthquakes are frequent, particularly among the most vulnerable populations.

The Jacobo Hunter district (as shown in Fig. 10.1A and B below) has an area of 2037 ha and a population of 46,216 inhabitants according to the 2007 census (INEI). Of the total district area, 615.32 ha (30.2%) are used for agricultural purposes, 484.66 ha (23.8%) correspond to the urban area; and 937.02 ha (46.0%) are desert, formed by hills (Municipality of Jacobo Hunter, 2007). Its gross density is 22.7 inhabitants/ha and the net density is 95.35 inhabitants/ha. The main hazards threatening the district include floods, earthquakes and landslides.

4 Development of PGIS model

There are several methodologies to identify the different phases of an emergency. The method proposed by the Peru National Disaster Risk Management Plan was chosen—PLANAGERD 2014–2021 (PCM, 2014), based on the "Hyogo Framework for Action 2005-2015." There are three components in the Disaster Risk Management: (a) Prospective Management (before the disaster takes place), (b) Reactive Management (the response after the crisis) and (c) Corrective Management (review of contingency plans). The project covers all stages of the DRM, but focuses on the reactive and corrective stage, which can serve as input for a new prospective stage. Fig. 10.2 below exhibits the three phases that are included in the proposed participatory GIS model.

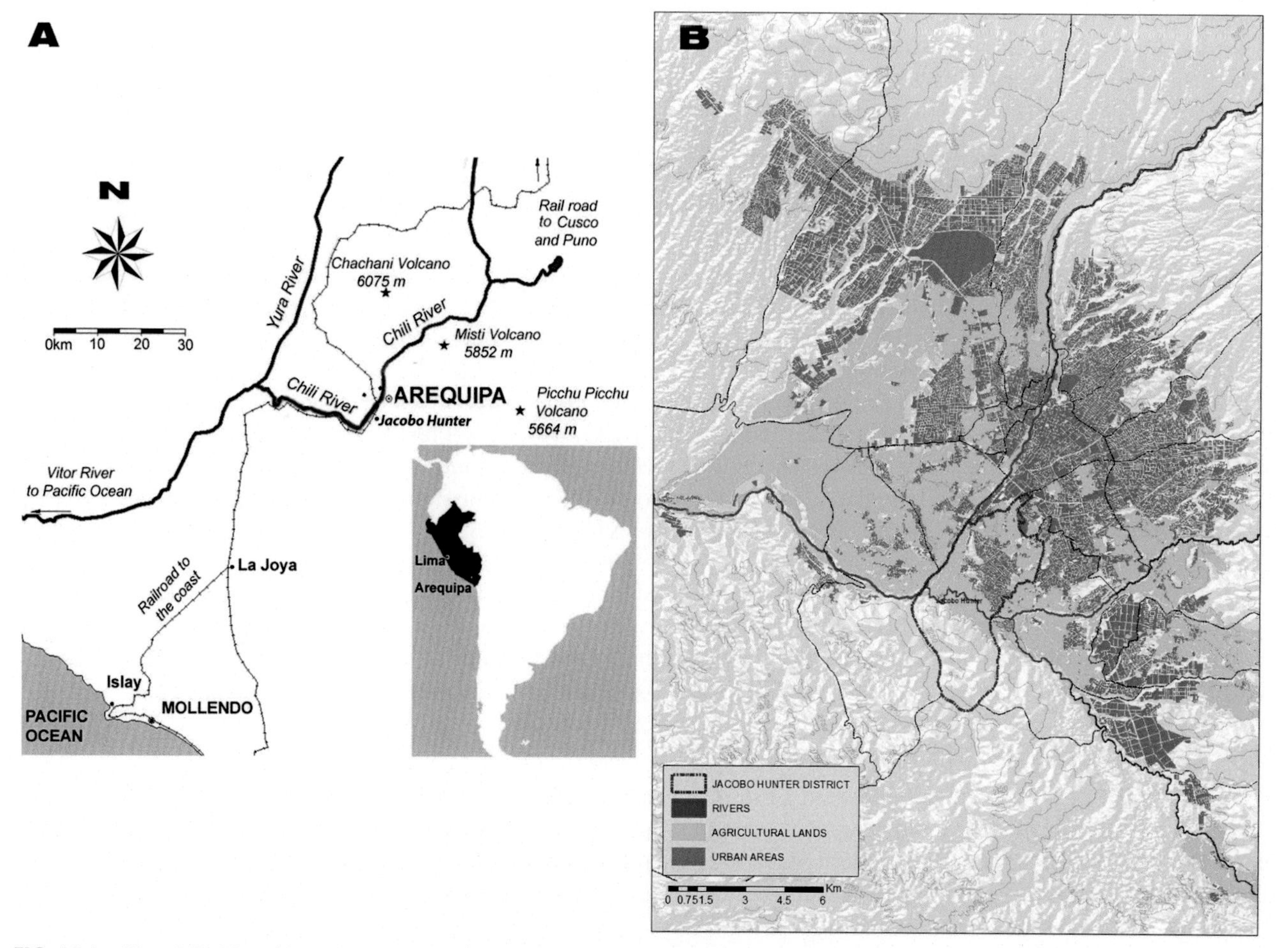

FIG. 10.1 (A) and (B) Map of Arequipa and Jacobo Hunter district.

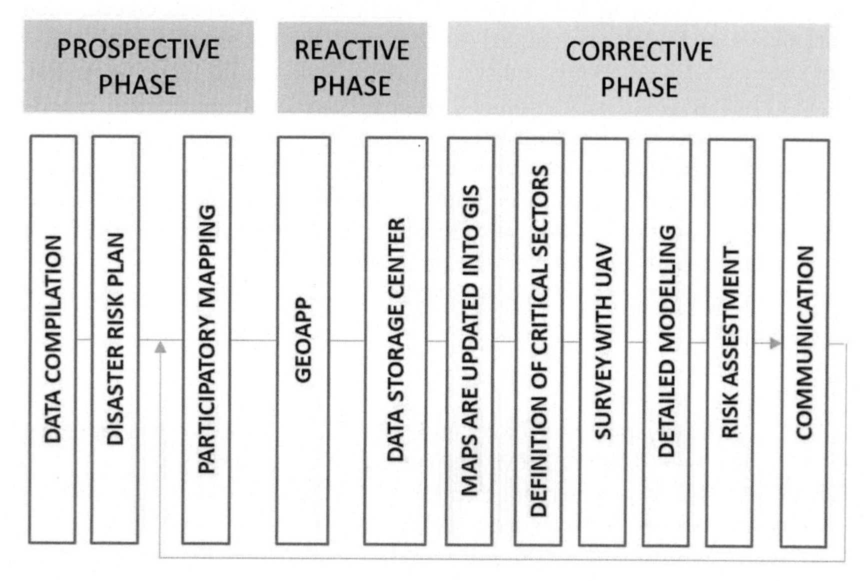

FIG. 10.2 Proposed participatory GIS model.

In developing the PGIS model, three types of main actors have been considered: (a) technicians, who are in charge of data collection, analysis and production of thematic maps in GIS; (b) organized population: who participates and gives their opinion about their experience with specific risks; (c) politicians and/or decision makers, who use the information generated in GIS to make decisions in the most affected areas.

4.1 Prospective management

Data compilation and analysis in GIS

For analysis of hazards, three threats were considered: (a) volcanic hazard, for which the hazard map of the National Center for Estimation, Prevention and Reduction of Disaster Risk, CENEPRED was consulted; (b) flooding hazard, taking data from the National Water Authority, ANA; and (c) seismic hazard, for which the study of the geological, geomorphological and seismic micro-zoning maps of the National University of San Agustín and the Civil Defence Institute INDECI was used. Additionally, areas of high and very high slope that are prone to landslides were determined, and a topography analysis was undertaken based on a Digital Elevation Model (DEM) with a pixel resolution of 30 m, which is a numerical representation of the surface of the earth based on a grid of specially aligned points, where each point represents a height of elevation. The DEM was obtained from the Advanced Spaceborne Thermal Emission and Reflection Radiometer (ASTER) satellite owned by NASA which is available on the United States Geological Survey (USGS) agency site (Municipality of Jacobo Hunter, 2016).

Data from the 2007 census projected for 2015 was also used for a vulnerability analysis in physical, social and economic aspects. Environmental vulnerability (UNDP, 2014) was not considered since the district will unlikely suffer from major pollution or deforestation problems. To calculate the vulnerability, a weighting associated with a numerical scale was developed according to the relevance of its physical and social components, obtaining a final map according to these weights.

Development of the district risk plan

The Municipality of Jacobo Hunter developed a risk plan where risk management policies are defined. However, a single multi-hazard map was produced, with the areas facing all three types of hazard being considered as "very high hazard" areas. The PGIS project, using the hazard information of the municipality and the moderated vulnerability value, included three hazard scenarios: flood, earthquakes and landslides.

4.2 Reactive management

Development of a geoapp

Given the occurrence of disaster events in recent years, the Disaster Risk Management (DRM) office of Arequipa is likely overwhelmed by the number of requests and their limited amount of resources to give a quick assistance (INDECI, 2015). In view of this situation, a geoapp or geographic-based mobile application has been developed, which allows the user to report the occurrence, magnitude, and location of natural (landslides, floods) or anthropogenic (fires, etc.) disasters. The application developed for smart mobile devices is called "Riesgo MApp" and was developed in Android Studio 3.0, using a real-time database of Google Firebase. The development of this app is based on an incremental, object-oriented model with the use of the Model View Controller (MVC) design pattern.

Two versions of the geoapp have been developed: (a) the user version, which allows the population to identify the place and the type and magnitude of an emergency by using descriptions and photos; (b) the administrator version, which allows receiving, classifying and viewing information from users, identifying which are the most affected areas and prioritizing more efficiently the use of limited resources to deal with the emergency.

4.3 Corrective management

A participatory workshop

A workshop was carried out in the Municipality of Hunter on Friday, March 17, 2017 with participation from communities and several institutions related to the district's DRM. The purpose of the workshop was to collect information such as (a) what were the main risks perceived by communities themselves, (b) the locations of these risks and (c) their intensity. During the workshop, community members were invited to locate, on printed maps and with colored markers, the various incidents that occurred in those areas, their type and magnitude. In the end, a discussion was held for people to openly talk about ways of improving risk management (Fig. 10.3).

FIG. 10.3 Participatory workshop.

Definition of critical sectors

After the workshop, the results of the participatory maps were collected and drawn in a georeferenced map, indicating the type, position and magnitude of the risks using the ArcMap software. Following the classification, a point density map was made to better identify the clusters of critical sectors, using the Kriging interpolation. The area identified as the most vulnerable is the *Buen Panorama* sector, characterized by a majority of precarious unplanned settlements on steep land without roads, public spaces or sufficient services.

Drone survey

After the critical sector was identified, a fixed-wing drone survey was carried out in an area of 2 ha. This survey was performed at a height of 120 m from the surface (See Fig. 10.4). The survey made in Pix4D allowed to obtain the following data:

- Orthomosaic in high definition: georeferenced aerial photography at a resolution of 5 cm per pixel, where many details can be seen in high quality.
- 3D modeling, which created based on a three-dimensional mesh to which an orthophotography was applied in to achieve visualization and bird-eye animation of the sector, using 3DS Max.
- Point clouds: a set of three-dimensional data that represent the external surface of objects.
- Contour curves: to represent the detailed topography of the study area.
- Digital Elevation Model (DEM): it is a representation of the surface in which risk events can be modeled, such as hydraulic flow, slope, among others.

Hydrological and slope modeling

Using the DEM surveyed by drone, two types of maps were developed in ArcMap:

- Slope analysis, classifying the slopes in low slope (< 8%), moderate (between 8% and 16%), high (between 16% and 35%) and very high slope (> 35%), according to Segura et al. (2011).
- Hydraulic runoff map, using hydraulic spatial analysis tools, such as flow direction, flow accumulation, and classification of currents according to the Strahler method, where "the classification of a stream it only increases when the streams of the same order intersect. Therefore, the intersection of a first order and second order link will maintain a second order link, instead of creating a third order link" (ESRI, 2016).

FIG. 10.4 3D model of the critical sector obtained by drone survey.

Calculating vulnerability in the critical sector

For the study of vulnerability in the critical sector, social, physical and economic vulnerability have been considered, according to the following criteria:

- Social Vulnerability Analysis: (a) Social Fragility has been taken into account, using population by plot as a variable and (b) Social Resilience, using the aptitude against risk as a variable.
- Physical Vulnerability Analysis: (a) Physical Fragility has been used, using as variables the predominant building materials, the height of the building and the status of the construction and (b) Physical Exposure, using as variables the provision of drinking water, drainage and electricity supply.
- Economic Vulnerability Analysis: Economic Resilience has been taken into account using variables such as family income and occupation of the head of the household.

The data was collected in surveys and field work (Cuzziramos, 2018). Subsequently, the criteria were assigned a differentiated weight according to the Process of Analytical Hierarchy (AHP), which is a structured technique that allows the selection, classification and prioritization of multi-criteria elements (Moreno-Jiménez, 2002). The method is based on the decomposition of complex structures in their primary elements, ordering them in a hierarchical structure, "where numerical values are obtained for preference judgments and, finally, synthesized to determine which variable has the highest priority" (Pacheco and Contreras, 2008). In this way it is possible to determine the vulnerability stratification which allows to indicate the criteria that were taken into account to determine each level.

Model validation

The final process included an in-situ workshop and a presentation of the results to the community. Using the detailed topography surveyed by drone, a physical model was made, accompanied by the maps of slopes and hydraulic flows, and it was possible to explain in a clear way the obvious risks to which the population is exposed. The aim of this final process was to solicit the users' opinions about whether the PGIS model is of use and how they possibly can mobilize it to attain the intended purpose.

5 Discussion and conclusions

Participatory mapping, as seen in the literature review, is an important way to empower people and include their experience as a relevant source for planning and decision making. Participatory Geographic Information Systems (PGIS) is an efficient combination of partaking approaches to planning and spatial information and communication management.

However, participation is not always easy or welcomed, particularly in Peru where both practice and legislation do not encourage the involvement of the communities in the Disaster Risk Management (DRM). Centralized, "black box" GIS carried out by technicians detached and far from the reality of hazardous zones as well as the need of spatial knowledge acquisition (Richardson et al., 1999) and cartographic illiteracy by the communities make participatory mapping a very uncommon practice.

The research reported in this chapter focuses on the importance of community participation in improving the effectiveness and efficiency of Disaster Risk Management, combining various tools of Geographic Information Systems and, particularly, on the ways both GIS technicians and vulnerable communities can benefit from each other to produce more accurate and reliable plans. Therefore, this research has expanded on the reviewed PGIS theory as participation can have various characteristics and particular approaches according to the stage of the DRM that is being worked on.

In a preventive stage, PGIS focused on data collection guided by the researchers to know from the community's own perspective which areas were most prone to the occurrence of disasters. Such insights did not necessarily coincide with those obtained by a "black box" GIS risk analysis. In this research, participatory workshops had allowed not only to know the locations of potential hazards but also their likely magnitude and recurrence likelihood. It was important to prepare the workshop in advance to facilitate communication and in that way get information from the community in a clear and easy way, and in a format in which such information could be readily processed and mapped using GIS in a later stage.

For the response stage, a PPGIS approach was used, as the community could use the developed mobile application as an efficient way to record and assess disaster events in real-time. The advantage of this method is that it allows mapping the concentration of incidence reports and decision-makers can prioritize the allocation of resources to the most severe emergencies. However, using such an application requires training for officials and community leaders, and availability of a smartphone for users, which is often not the case in economically disadvantaged communities, who might face the greatest risks and be most in need after an event.

For the corrective stage, deeper analyses were made in the sectors previously identified as critical by the population. A survey using a drone in these sectors allows a much more accurate information gathering without incurring an excessive increase in costs. The detailed information allows identifying the built and vacant areas, vegetation, materials, volumes and, in the case of hazardous areas, such as prone to floods and landslides.

Likewise, more emphasis must be placed on the determination of vulnerability, since it is the factor in which more control can be done to mitigate the risks of a disaster. However, it is necessary for the vulnerability analysis to include social, physical, economic aspects and, if applicable, an environmental approach. For the composition of these vulnerability aspects, the Analytical Hierarchy Process was proposed. A major limitation in this regard is the lack of social data at the block level, and this gap might be filled by using surveys or interviews and fieldwork in the future.

Finally, and in light of the results obtained in the district of Jacobo Hunter, this method can serve to establish a dialogue with local communities, who would be typically reluctant to listen to politicians or municipal authorities especially when dealing with issues such as the relocation of housing in areas facing not-mitigable risk. Such a PGIS model can demonstrate the effects of a disaster in vulnerable areas in a user friendly manner.

It is suggested that the future research can include testing and validating this methodology in other neighboring districts, and eventually at the metropolitan level. For this purpose, adjustments at the logistic level should be taken into account to guarantee the participation of communities and the management of information on a larger scale. A future research project can also develop a functional prototype of participatory risk management for the peripheral areas of the city of Arequipa, as well as the measurement of its effectiveness.

Acknowledgments

The author wish to express their gratitude to the Vice-Rector for Research of the Universidad Católica de Santa María, Dr. Gonzalo Dávila del Carpio for sponsoring the "Participatory Geographic Information Systems for Integrated Risk Management in Vulnerable Populations of the Periphery of Arequipa" research project, to which this research is affiliated. I also appreciate the important collaboration of Gustavo Delgado Alvarado, Fernando Cuzziramos Gutiérrez and Verónica Salinas Murillo, for their participation in the workshops, as well as for their ideas that helped in the discussion. Also, the collaboration of professors Sergio Poco Aguilar and José A. Herrera Quispe and students Vidal Soncco Merma and Diego Ranilla Gallegos, who collaborated in the development of the geoapp.

In addition, I wish to thank Mr. Simón Balbuena Marroquín, Mayor of the Municipality of Jacobo Hunter, for his support in summoning the population, providing data and arranging the collaboration of the municipal officials.

References

Browna, G., Kyttä, M., 2018. Key issues and priorities in participatory mapping: toward integration or increased specialization? Appl. Geogr. 95, 1–8. https://doi.org/10.1016/j.apgeog.2018.04.002.

CENEPRED, 2014. Manual de Evaluación de Riesgos de Desastres por Fenómenos Naturales. Centro Nacional de Prevención y Reducción de Desastres, Lima.

Chambers, R., 2006. Participatory mapping and geographic information systems: whose map? Who is empowered and who disempowered? Who gains and who loses? Electron. J. Inf. Syst. Dev. Countries 25 (2), 1–11.

Chingombe, W., Pedzisai, E., Manatsa, D., Mukwada, G., Taru, P., 2015. A participatory approach in GIS data collection for flood risk management, Muzarabani district, Zimbabwe. Arab. J. Geosci. 8 (2), 1029–1040. https://doi.org/10.1007/s12517-014-1265-6.

Cuzziramos, F., 2018. La Vivienda Colectiva Como Alternativa Ante la Gestión Integrada de Riesgos en Poblaciones Vulnerables. Propuesta de Intervención en el Distrito de Jacobo Hunter, Arequipa (Master thesis). Universidad Católica de Santa María, Arequipa.

ESRI, 2016. Clasificación de Corrientes. ArcMap. Retrieved from: http://desktop.arcgis.com/es/arcmap/10.3/tools/spatial-analyst-toolbox/stream-order.htm.

Giovene di Girasole, E., Cannatella, D., 2017. Social vulnerability to natural hazards in urban systems. An application in Santo Domingo (Dominican Republic). Sustainability 9 (11), 2043. https://doi.org/10.3390/su9112043.

Grupo GEA, 2008. Guía Metodológica para el Ordenamiento Territorial y la Gestión de Riesgos. Grupo GEA, Lima.

IFAD, 2009. Good Practices in Participatory Mapping. International Fund for Agricultural Development. Retrieved from: https://www.ifad.org/documents/38714170/39144386/PM_web.pdf/7c1eda69-8205-4c31-8912-3c25d6f90055.

INDECI, M., 2015. Situación de la Oficina de Defensa Civil Arequipa ante los Desastres Naturales. (C. Zeballos, Interviewer, marzo 05).

McCall, M.K., 2014. Mapping territories, land resources and rights: communities deploying participatory mapping/PGIS in Latin America. Rev. Depart. Geografia, 94–122. Volume Especial Cartogeo.

Ministry of Environment, Peru, 2015. Guía Metodológica Para la Elaboración de los Instrumentos Técnicos Sustentatorios Para el Ordenamiento Territorial. MINAM, Lima.

Moreno-Jiménez, J., 2002. El Proceso Analítico Jerárquico (AHP). Fundamentos, Metodología y Aplicaciones. Dpto. Métodos Estadísticos. Facultad de Económicas. Universidad de Zaragoza. Retrieved from: https://users.dcc.uchile.cl/~nbaloian/DSS-DCC/ExplicacionMetodoAHP(ve%20rpaginas11-16).pdf.

Muller, M., 2007. Adapting to climate change: water management for urban resilience. Environ. Urban. 19, 99–113.

Municipality of Jacobo Hunter, 2007. Plan de Desarrollo Concertado 2007–2015. Municipalidad Distrital de Jacobo Hunter, Arequipa.

Municipality of Jacobo Hunter, 2016. Plan Urbano Distrital 2016–2025. Municipalidad Distrital de Jacobo Hunter, Arequipa.

Pacheco, J., Contreras, E., 2008. Manual Metodológico de Evaluación Multicriterio para Programas y Proyectos. Instituto Latinoamericano y del Caribe de Planificación Económica y Social (ILPES), Santiago de Chile.

PCM, 2014. Plan Nacional de Gestión del Riesgo de Desastres—PLANAGERD 2014–2021. Presidencia del Consejo de Ministros, Lima.

Radil, S.M., Jiao, J., 2016. Public participatory GIS and the geography of inclusion. Prof. Geogr. 68 (2), 202–210. https://doi.org/10.1080/00330124.2015.1054750.

Rambaldi, G., Kwaku Kyem, P.A., McCall, M., Weiner, D., 2006. Participatory spatial information management and communication in developing countries. Electron. J. Inf. Syst. Dev. Countries 25 (1), 1–9.

Richardson, A.E., Montello, D.R., Hegarty, M., 1999. Spatial knowledge acquisition from maps and from navigation in real and virtual environments. Mem. Cogn. 27 (4), 741–750.

Segura, G., Badilla, E., Obando, L., 2011. Susceptibilidad al deslizamento en el corredor Siquirres-Turriba. Revista Geológica de América Central 45, 101–121.

Tane, H., Wang, X., 2007. Participatory GIS for sustainable development projects. In: 19th Annual Colloquium of the Spatial Information Research Centre University of Otago, Dunedin, New Zealand.

UNDP, 2014. Cuba. Metodologías Para la Determinación de Riesgo de Desastres a Nivel Territorial. Programa de Naciones Unidas para el Desarrollo, La Habana.

UNISDR (United Nations Office for Disaster Risk Reduction), 2009. Terminology on Disaster Risk Reduction. Retrieved 06 16, 2020, from: https://www.preventionweb.net/files/7817_UNISDRTerminologyEnglish.pdf.

Chapter 11

Social capital in disaster recovery: A case study after the 2016 earthquake in Ecuador

Laura Cevallos-Merki[a] and Jonas Joerin[b]

[a]*Department of Geography, University of Zurich, Zürich, Switzerland,* [b]*Singapore-ETH Centre, Future Resilient Systems, Singapore, Singapore*

1 Introduction

In recent years, Disaster Risk Management (DRM) has undergone a shift from looking only at physical aspects toward a more integrated approach, including also social aspects. The recovery phase after a disaster is seen as an opportunity for development and risk reduction by "building back better." It is nowadays agreed that, in order to fulfill these demands and meet the community needs, it is crucial to include the community in the recovery process. That is, on one hand consider social aspects such as culture and power relations in the recovery process and on the other hand work with the community in the planning as well as the implementation phase. As Nakagawa and Shaw (2004, p. 12) conclude from their study of the recovery process after the earthquakes in Kobe, Japan and Guajarat, India, "*it was the community which determined whether each member was satisfied by the rehabilitation*." Therefore it is also important to apply bottom-up recovery policies.

Due to this trend, a discussion has started about the role of Social Capital (SC) in disaster recovery. SC can be defined as a function of "*trust, social norms, participation and networks*" (Nakagawa and Shaw, 2004, p. 1).

Different researchers have shown that communities with high SC generally experience a quicker and more sustainable recovery and higher recovery satisfaction than communities with low SC (Joshi and Aoki, 2014; Nakagawa and Shaw, 2004; Sanyal and Routray, 2016). However, still little research exists in this area. Our aim therefore was to expand findings on the role of SC in disaster recovery. More precisely we investigated how SC influences recovery satisfaction and which networks played a major role in the recovery process. In contrast to other studies we did not measure the physical state of recovery but only looked at people's satisfaction with the recovery process.

Studies about the role of SC in disaster recovery were conducted in India, Japan and East Azerbaijan (Alipour et al., 2015; George, 2008; Joshi and Aoki, 2014; Nakagawa and Shaw, 2004; Sanyal and Routray, 2016), but no such study was found in Latin America or Ecuador. As a case study we therefore chose the recovery process after the 2016 earthquake in the cantons of Jama and Pedernales of Ecuador. The earthquake hit the coastal area of Ecuador in April 2016, with a magnitude of 7.8 on the Richter scale and left 720,000 people in need of help (Gobierno de Ecuador, 2016; OCHA, 2016; Telesur, 2016). Different stakeholders were part of the recovery process, though the national government led and coordinated the process (SGR, 2014, 2016a,b; SNGR, 2012). The most important policy that led the recovery process was the national government's reconstruction plan (Reconstruyo Ecuador), which followed a rather top-down recovery process: The government provided mostly physical help in form of houses and food but little did it promote community inclusion and workshops.

2 Literature review

DRM is a complex topic, as many different dimensions and stakeholders are impacted by or have an influence on managing disasters. Various sectors (e.g., politics, economy, city planning, culture), different geographical units (local, regional, national, international) and different actors on a community level are part of a disaster recovery. The characteristics of these sectors and actors can vary greatly between cases of disaster recovery. Flexibility toward local social conditions, such as culture and power relations, is therefore crucial in DRM. Disasters can neither be prevented nor properly recovered if the social aspects are not fully taken into account (Eiser et al., 2012).

Due to the increasing importance of social aspects in DRM a discussion has started about the role of SC in disaster recovery. SC can foster many elements of a sustainable long-term recovery such as collective action, mutual support and community participation (Joshi and Aoki, 2014; Nakagawa and Shaw, 2004; Sanyal and Routray, 2016).

Strengthening Disaster Risk Governance to Manage Disaster Risk. https://doi.org/10.1016/B978-0-12-818750-0.00011-8

2.1 Social capital

A wide variety of literature exists about SC. Capitals in general are tools, which facilitate the production of other resources or future activity (Coleman, 1988; Ostrom, 2000). Different kinds of capital are never totally independent. An actor's networks (social capital) can for example be greatly influenced by its physical assets (physical capital) and education (human capital). The SC concept originally emerged in social science as an addition to physical and human capital. It later spread to political and to economic sciences and has in recent years gained importance in development sciences, and also in DRM literature (Grootaert et al., 2004; Hanifan, 1916; Uphoff, 2000; Woolcock, 1998). SC is not a single entity but consists of various functions. Therefore, depending on the researcher's focus, different definitions of SC are used. However, it is generally agreed that SC consists of some kind of social structure, which fosters collective action or facilitates actors to access certain resources (Portes, 1998; Putnam, 2001; Woolcock, 1998). This paper works with the definition of SC by the functions that include *social networks*, *trust* and *norms* (Joshi and Aoki, 2014; Nakagawa and Shaw, 2004; Serageldin and Grootaert, 1999). These functions are interconnected and can influence each other to a high extent.

Social networks are formal or informal relationships between the members of a group characterized through social exchange and interactions and held together through the norms of reciprocity (Putnam et al., 1993; Uphoff, 2000). Some disagreement exists on whether strong or weak ties within networks lead to higher SC. Strong ties foster information flow and trust between members within a network. However, it can also lead to closure toward outsiders of the network. Weak ties can foster exchange of information and help between networks and therefore lead to advantages for members of both networks (Burt, 2017; Coleman, 1988; Putnam et al., 1993). *Trust* between members of a group is an important part of SC because cooperation among people and thus collective action would not be possible without it (Putnam et al., 1993). *Social norms* exist in different forms, such as obligations and the willingness towards mutually beneficial action and reciprocity. They can be formal or informal (Nakagawa and Shaw, 2004; Putnam et al., 1993).

Different categorizations of SC exist. Researchers differentiate between Horizontal and Vertical SC, Bonding, Bridging and Linking SC, Individual and Community SC, Formal and Informal SC and Structural and Cognitive SC. However, this paper only differentiates between Vertical and Horizontal SC. Horizontal SC is further divided into Individual and Community SC. While Vertical SC consists of networks with asymmetric power relations, for example, relations between community members and authority figures, Horizontal SC consists of networks between people of the same hierarchical level (Grootaert, 2001; Putnam et al., 1993). Individual SC focuses on social relationships that enable individuals or households to get access to resources, while Community SC describes the interaction and collaboration of people within a community, which provide an advantage for all members of the community (Dudwick et al., 2006; Grootaert et al., 2004; Son and Lin, 2008).

2.2 Social capital in disaster recovery

Different studies have shown that including SC in the recovery process is vital for a sustainable long-term recovery (Joshi and Aoki, 2014; Nakagawa and Shaw, 2004; Minamato, 2010; Sanyal and Routray, 2016). Communities with high SC have reported, in general, greater recovery satisfaction and greater community participation during the recovery process and overall faster and more successful recovery results after a disaster (Joshi and Aoki, 2014; Nakagawa and Shaw, 2004; Sanyal and Routray, 2016). Including SC in the recovery process is especially important as disasters can entail serious disruptions in SC, while recovery policies can either activate or undermine SC.

SC can be included in the assessment after a disaster as well as in the planning and implementation phase. SC as a concept helps to understand and analyze community structures and thus to find appropriate recovery measures. Including SC in the planning phase can lead to higher acceptance of the recovery policies. SC as a resource can foster collective action and community participation during the planning as well as the implementation phase. It can facilitate people to access aid and resources such as getting assistance from others for childcare, shelter and emotional support (Joshi and Aoki, 2014; Nakagawa and Shaw, 2004; Sanyal and Routray, 2016). Nakagawa and Shaw (2004) for instance examined the role of SC in the reconstruction effort in Kobe, Japan and in Gujarat, India after the respective earthquakes. They observed that communities with higher levels of SC recovered faster and in a more sustainable way than communities with weaker SC.

3 Methods

In the course of this study several methods were used to obtain and analyze data. The main results were acquired by a quantitative household survey. To complement the results of the household survey, field notes, notes from informal conversations and supplementary statements from the survey were recorded and analyzed.

TABLE 11.1 Data aggregation: Topics that belong to the respective SC classes with the according methods used for data aggregation.

New variable	Topic	Actors
Individual SC	Number of relations Number of reunions Trust	Family Friends Neighbors
Community SC	Trust Security Mutuality	People of the neighborhood People of the village
CBOs	Network with CBOs	CBOs
Vertical SC	Trust Access	Neighborhood president Local council (cantonal level) National government NGOs
Collective action and participation	Collective action Participation	Measured on a community level

The surveys were conducted between January and March 2017 (9–11 months after the earthquake) in the cantons of Jama and Pedernales, province of Manabí, Ecuador. The province of Manabí was chosen because it was among the most affected provinces by the 2016 earthquake (Oxfam, 2016). In total, 203 households participated in the survey. The survey participants were visited at their houses or in the camps and were guided through the questionnaire.

The samples were selected randomly with some considerations: Jama and Pedernales should have the same number of samples and questionnaires should be equally distributed among gender. Also people living in reception camps, which were built after the earthquake, should be included. However, the random sampling was limited to some points, as the entrance into very poor and dangerous neighborhoods was difficult.

The questionnaire contained only closed questions in form of multiple-choice questions. It was divided into four main sections: Demographic and regional data; Damage; Satisfaction with the recovery so far; and, SC. Questions about SC were formulated based on guidelines from the World Bank: Measuring Social Capital—An Integrated Questionnaire (Grootaert et al., 2004) and Instruments of the Social Capital Assessment Tool (Grootaert and Van Bastelaer, 2002).

Different questions concerning SC were aggregated into classes: Individual, Community and Vertical SC. The aggregation was done by adding the scores of the respective questions. In some cases, a weighted sum was used, i.e., when one topic or one question was considered more important than others. As little Community-Based Organizations (CBOs) existed in the survey area the membership of a CBO was treated as a separate class and not included in Community SC. Additionally a category concerning collective action and participation on a community level during the recovery was created (Table 11.1). The IBM program Statistical Package for the Social Sciences (SPSS version 22.0.0.0) was used to get statistical data such as mean, median and distributions, to find possible differences between the cantons (Kruskal-Wallis-Test and Mann-Whitney-U-Test) and to find possible correlations between SC and recovery satisfaction (Kendall's Tau-b-Test).

Qualitative data was obtained through conversations with affected people that were held during the survey. They were recorded in a field book, as well as key messages were written in an additional box that was part of the questionnaire.

4 Results

4.1 Case study site

The cantons Pedernales and Jama are both located in the north-western region of the Ecuadorian coast, in the province of Manabí. The canton of Pedernales has 61,000 inhabitants and a poverty rate of 94%. The canton of Jama counts 25,400 inhabitants and a poverty rate of 90%. The main employment sectors in both cantons are agriculture, farming, fishing and forestry.

Ecuador has a decentralized system of risk management (SGR, 2014, 2016a,b; SNGR, 2012). Different ministries and institutions play a role in disaster risk management (e.g., the ministry for economic and social inclusion (MIES), the ministry for development and housing (MIDUVI), etc.). DRM is therefore the responsibility of each institution in its sectorial and geographical area (SGR, 2016a,b). However, the state has the exclusive competency over natural disasters (SGR, 2016a,b). After the 2016 earthquake Ecuador also received help from the United Nations. An important part of the government's recovery projects was the plan "Reconstruyo Ecuador" (Reconstruct Ecuador). When people were asked about the reconstruction, every single person started talking almost exclusively about "Reconstruyo Ecuador." This plan implies that every person who has lost her or his house, whose house is located in a risk prone zone or who has rented a house in such a zone, can apply for a new house, which would be up to 90% donated by the government (MIDUVI, 2016).

4.2 State of recovery and recovery satisfaction

The description of the state of recovery is based on observations and qualitative statements gathered during the stay in the research area. At the time of the data collection (9–11 months after the earthquake) the temporary camps and shelters were still full, as only few houses have been rebuilt. However, most of the damaged streets have been repaired and the schools and the local council were functioning in temporary shelters.

The results about the recovery satisfaction are based on a quantitative survey. A distinction was made between satisfaction with the house recovery, with mental health recovery and with the village recovery. People could choose between five options from very satisfied, satisfied, indifferent, dissatisfied to very dissatisfied. Satisfaction with house recovery was low, with the median of answers concentrated on the *dissatisfied* category. Satisfaction with the village recovery varied greatly within the population and scored a median of *indifferent*. However, people were generally satisfied with their mental health recovery (median *satisfied*).

4.3 Social capital in Jama and Pedernales

Fig. 11.1 shows that Individual SC (average 12.24) is rather high in Jama and Pedernales. Community SC is close to Individual SC with an average of 11.21. However, it is important to take into account that the membership in a CBO is not included in this rating, as hardly any CBOs existed in the area. Vertical SC is very low with an average of 4.48. Recovery participation scored an average of 8.

In the following subchapters these differences are discussed in more detail.

4.4 The function of trust

The function of trust was measured on a five item scale, with categories from totally, a lot, somewhat, a little to nothing at all. Trust in individual networks was generally rated high. This accounts especially for trust in other family members, as people stated that they could trust their family *totally* (Table 11.2). Even though Community SC scored relatively well, trust in other community members was rated low with only *a little* (Table 11.2). Trust into authority figures was overall rated low. However, people have more trust in the national government (*a little*) than in the neighborhood president and the local council (*nothing at all*). Of all authority figures, NGOs scored best with an average trust of *somewhat*.

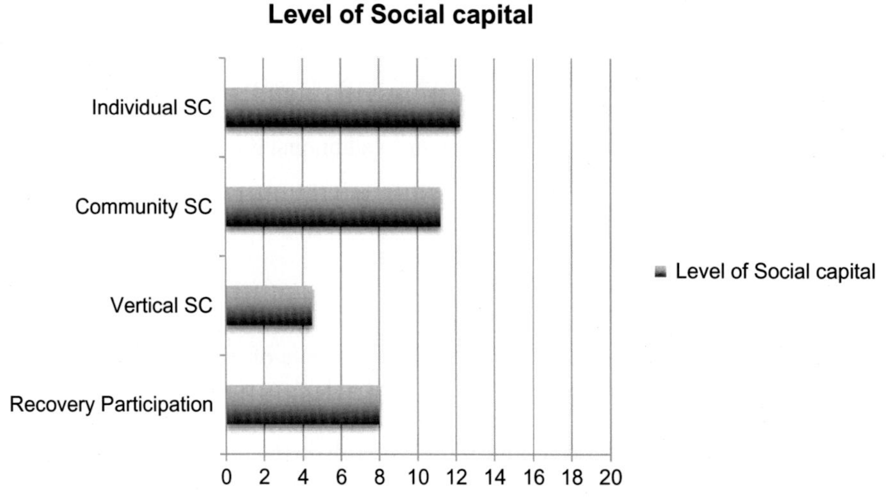

FIG. 11.1 Level of SC in Jama and Pedernales ranging from 0 to 20, responding to the questions from the questionnaire (see Section 3).

TABLE 11.2 Median of trust to different networks.

Type of network	Subcategory	Median
Individual trust	Family	Totally (1)
	Friends	A lot (2)
	Direct neighbors	Somewhat (3)
Community trust	Neighborhood	A little (4)
	Village	A little (4)
Vertical trust	Neighborhood president	Nothing at all (5)
	Local council	Nothing at all (5)
	National government	A little (4)
	NGOs	Somewhat (3)

4.5 The function of norms

The function of norms was measured through the variables "feeling of security" and "willingness to participate in community activities." It was not possible to measure the habit of democratic decision-making, as people were not used to be part of decision-making and therefore did not understand the questions we asked.

People felt *moderately safe* in the street and *neither safe nor unsafe* at home. Many people stated that they would *surely participate* in a community activity to do some recovery work for their community (median *yes for sure*). This is in contrast to the experience that in practice only few people participated in a recovery activity.

4.6 The function of networks

In order to measure the function of networks in the two communities, we investigated how many people were part of a formal or informal organization, association or group. We found that hardly any Community-Based Organizations (CBOs) existed in the area. Less than 25% of all surveyed persons were part of an organization, association or group and even less stated that they got help from an organization they belong to during the recovery.

Additionally, the question was posed of which networks were most active during the recovery process. Thereby, a difference was made between the variables "sources of help" and "networks that worked together."

People stated that vertical networks (mainly the national government and NGOs) provided most help during the recovery process (Fig. 11.2). However, they did not experience much collaboration with vertical networks (Fig. 11.3). Individual networks (mainly family) were an important source of help too (Fig. 11.2). Also, people stated that they collaborated very closely within individual networks (Fig. 11.3). Community networks worked together to a small extent during the recovery process. This accounts mainly for people of the village. CBOs however, played a minor role.

4.7 The function of collective action and recovery participation

Many people said that they were *very willing to participate* in a community activity for the recovery of the village. However, out of 203 surveyed people only 92 stated that they had participated in a communal activity for the recovery after the earthquake. Most of them participated in short-term recovery activities such as *clearance of debris* (named 42 times) followed by *building houses* (named 24 times), *education & health* (named 17 times) and *repairing infrastructure* (named 14 times).

4.8 Correlation between SC and recovery satisfaction

Considering the characteristics of our variables (non-parametric, ordinal) and the sample size (200), Kendall's tau-test was used to analyze the correlation between SC and recovery satisfaction. Our results using Kendall's-tau-test (Table 11.3) show that satisfaction with house recovery correlates significantly with Individual SC ($P=.002$; tau $=-0.162$), Community SC ($P=.005$; tau $=-0.149$) and Vertical SC ($P=.001$; tau $=-0.174$). Satisfaction with mental health recovery correlates

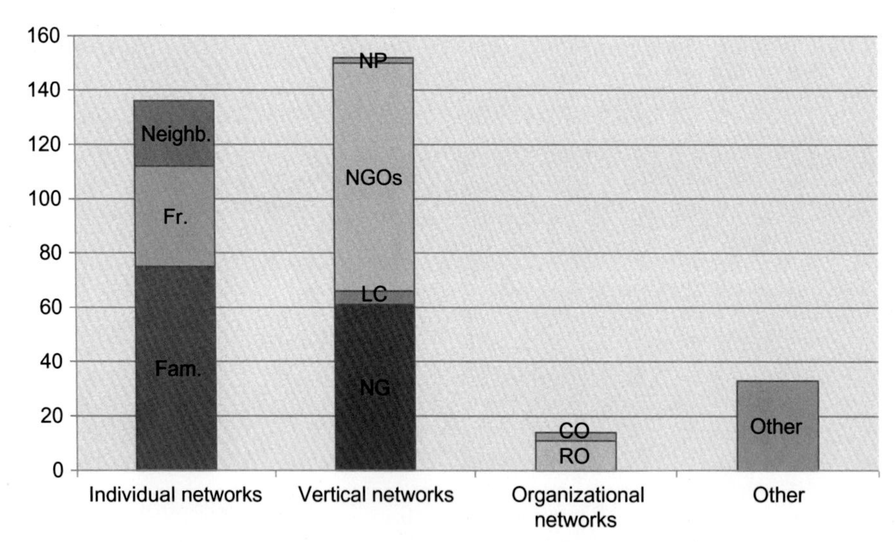

FIG. 11.2 Actors that helped most in the earthquake recovery aggregated into Individual, Vertical and Organizational networks and Other (various) (surveyed persons could choose between 1 and 3 actors) ($n=203$). *Fam.*, family; *Fr.*, friends; *Neighb.*, neighbors; *NG*, national government; *LC*, local council; *NP*, neighborhood president; *RO*, religious organizations; *CO*, civic organizations.

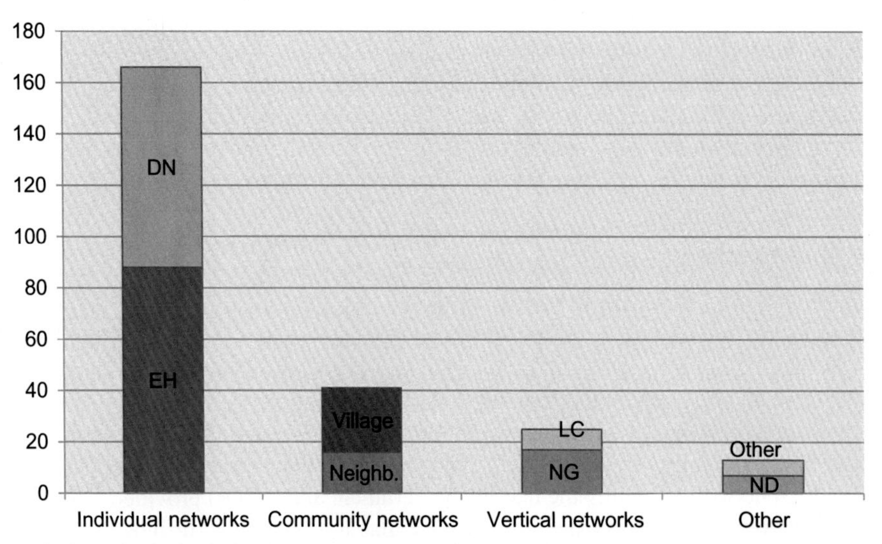

FIG. 11.3 Actors that worked together in the earthquake recovery aggregated into Individual, Vertical and Organizational networks and Other (various) (surveyed persons could choose several actors) ($n=203$). *EH*, each household; *DN*, direct neighbors; *Neighb.*, neighborhood; *NG*, national government; *LC*, local council; *ND*, no damage.

significantly with Vertical SC ($P=.033$; tau $=-0.155$). Satisfaction with the village recovery correlates significantly with Vertical SC ($P=.000$; tau $=-0.224$), as well as with collective action ($P=.033$; tau $=-0.116$) (Table 11.3). The direction of the correlation is negative, which in this case means that high SC leads to greater satisfaction. The results show that especially Vertical SC (trust and access to authority figures) has a positive influence on the recovery satisfaction, while Community SC only correlates with the satisfaction with house recovery.

4.9 Qualitative results

Conversations with the people living in the affected area (participants of the household surveys as well as people in the streets and authority people such as police men, soldiers and people working for the local government) confirmed the findings that individual networks were, after the vertical networks, the ones that helped the most in the long-term earthquake recovery. People explained that they trusted their families more than their friends and neighbors. Especially in the first instance after the earthquake it were mostly family networks that helped to safe lives. To give an example of one of these

TABLE 11.3 Correlation between SC and Recovery satisfaction (Kendall-Tau-b-Test) ($n=200$).

	Satisfaction with house recovery	Satisfaction with mental health recovery	Satisfaction with village recovery
Individual SC	0.002**	0.456	0.338
Community SC	0.005**	0.194	0.738
Membership of CBO	0.514	0.683	0.138
Vertical SC	0.001**	0.033*	0.000***
Collective action and participation	0.371	0.907	0.033*

*$P<.05$.
** $P<.01$.
*** $P<.001$.

statements: "*Our neighbor died. We heard her scream for help [...] Why we did not help? You can not imagine an earthquake! It was a disgrace! Everyone took care of his family and left.*" (translated from Spanish into English after LM 158). Also a police officer who was talking about the recovery process explained: "*The family always comes first.*" (translated from Spanish into English after P6). This shows how the ties of family networks are strong.

Due to the reconstruction plan, people had high expectations of the government. However, several conflicting issues were found with the recovery plan around 10 months after the earthquake: people claimed that (1) many houses were still not allotted; (2) houses got allotted when they were only halfway finished and (3) the distribution of houses was unfair and (4) they did not take action, as they preferred to wait for the government's promised aid. This resulted in many people playing a passive role in the recovery process and high dissatisfaction about the overall recovery process. Also, a lot of distrust and envy had evolved within communities as people stated that some families received more help than others: "*MIDUVI gives the houses to people who already have enough!*" (translated from Spanish into English after LM 191) "*They distributed the food only among themselves* (the family networks and close friends) *and therefore it went bad.*" However, some people also stated that trust and networks among the people of the village had increased: "*The people are a lot more united now than they were before.*" (translated from Spanish to English after LM 72).

5 Discussion

At the time of collecting the data in Jama and Pedernales, around 9–11 months after the earthquake, recovery was still not advanced and people's satisfaction with the state of recovery was low. This accounts especially for the satisfaction with the house recovery and village recovery.

Our results show that Vertical SC correlated with each subdivision of recovery satisfaction. People who strongly trust and had easy access to authority figures were more satisfied with the state of recovery. A very strong positive correlation was found between Vertical SC and satisfaction with village recovery, as also a strong correlation between Vertical SC and satisfaction with house recovery. Individual SC correlated only with the satisfaction in house recovery. Accordingly, people with strong trust in family members, friends and neighbors and who were part of strong individual networks felt more satisfied with their house recovery. Community SC only correlated with the satisfaction with house recovery. People who rated community norms strong and who felt strong trust to other community members felt more satisfied with the house recovery. Membership of a CBO did not correlate with recovery satisfaction at all, while collective action and participation only showed a weak correlation with village recovery satisfaction. We assume that people with strong community networks were more commonly part of community activities and therefore felt more satisfied with their house reconstruction. However, it is surprising that Community SC did not correlate with village recovery and neither did Membership of a CBO.

That recovery satisfaction correlates mostly with Vertical SC is in contrast to findings of other studies, which found community networks to be the main influence on recovery perception and satisfaction. Minamato's (2010) results, for example, show that formal community networks such as CBOs are the factors which best describe livelihood recovery perception. However, in Jama and Pedernales membership of CBOs did not have any influence on recovery satisfaction. We assume that the difference in findings can be traced back to a lack of CBOs, weak community trust before the earthquake and a top-down recovery approach in Jama and Pedernales.

The importance of Vertical SC for the recovery satisfaction may be traced back to the government's recovery approach, which mainly provided aid in form of physical measures such as food, shelter and houses in order to help affected communities to recover from the earthquake. Yet, people did not feel a strong cooperation with authority figures, as the recovery approach was rather top-down with little local inclusion. People especially stressed that local authority figures such as the local council and neighborhood presidents did not support them during the recovery process and neither did they collaborate with the community. Trust in local authority figures was very low. This might have been a major drawback in the recovery process of Jama and Pedernales as other studies showed, that trust in local leaders can lead to enhanced collective action on a community level and better recovery results (Joshi and Aoki, 2014; Nakagawa and Shaw, 2004).

Individual networks played an overall important role in the recovery process. A lot of collaboration within individual networks was observed, especially within families. This may be traced back to strong trust among family members. However, our results show that strong family ties also led to closure toward outsiders in some cases. In this point we agree with Minamato (2010) who found that tight family networks lead in some cases to lower recovery satisfaction. It is therefore especially important to distribute aid equally in order to prevent envy.

In Jama and Pedernales community networks played only a minor role in the recovery process. This is surprising, as Community SC was rated higher than Vertical SC. However, hardly any CBOs existed in the area. Therefore people were not used to organizing themselves. Weak community trust may also have had a negative influence on community collaboration. The government's top-down recovery approach additionally undermined community activities. It also fostered closure between family networks, as in some cases the distribution of aid led to envy.

In summary it can be said that collaboration was mainly observed among family networks and to some extent among friends and direct neighbors. Some community activities to recover after the earthquake were observed but most of them only served the short-term recovery. Very tight family networks, distrust among community members, a lack of organization, weak local leadership and the government's top-down approach inhibited collective community recovery. Other studies have found that strong community networks, that is CBOs and strong community trust, strong local leadership and bottom-up recovery plans can lead to enhanced collective action and better overall recovery satisfaction. We therefore recommend strengthening community trust and organization through enhancing CBOs and through organizing workshops on a community levels. This may also lead to stronger networks between families. We also recommend implementing recovery policies with more community inclusion and give more power to the local council, which is the intermediary between the community and the national government. It is thereby important to include the local council in community workshops in order to strengthen networks and trust between the community and the local authority figures.

References

Alipour, F., Khankeh, H., Fkrazad, H., Kamali, M., Rafiey, H., Ahmadi, S., 2015. Social issues and post-disaster recovery: a qualitative study in an Iranian context. Int. Soc. Work. 58 (5), 689–703.

Burt, R.S., 2017. Structural holes versus network closure as social capital. In: Social Capital. Routledge, pp. 31–56.

Coleman, J.S., 1988. Social capital in the creation of human capital. Am. J. Sociol. 94, 95–120.

Dudwick, N., Kuehnast, K., Nyhan Jones, V., Woolcock, M., 2006. Analyzing Social Capital in Context: A Guide to Using Qualitative Methods and Data. World Bank, Washington, DC. Retrieved 18.06.2020, from: http://documents.worldbank.org/curated/en/601831468338476652/Analyzing-social-capital-in-context-a-guide-to-using-qualitative-methods-and-data.

Eiser, J.R., Bostrom, A., Burton, I., Johnston, D., McClure, J., Paton, D., van der Pligt, J., White, M., 2012. Risk interpretation and action : a conceptual framework for responses to natural hazards. Int. J. Disaster Risk Reduct. 1, 5–16.

George, B.P., 2008. Local community's support for post-tsunami recovery efforts in an agrarian village and a tourist estination: a comparative analysis. Community Dev. J. 43 (4), 444–458.

Gobierno de Ecuador, 2016. PDNA—Reconstrucción y Reactivación de las Zonas Afectadas por el Terremoto del 16 de Abril Ecuador. Retrieved 18.06.2020, from: https://www.humanitarianresponse.info/fr/operations/ecuador/document/reconstrucci%C3%B3n-y-reactivaci%C3%B3n-de-las-zonas-afectadas-por-el-terremoto.

Grootaert, C., 2001. LLI 10—Does Social Capital Help the Poor? A Synthesis of Findings From the Local Level Institutions Studies in Bolivia, Burkina Faso and Indonesia. World Bank, Washington, DC. Retrieved 12.10.2016, from: http://siteresources.worldbank.org/INTRANETSOCIALDEVELOPMENT/882042-1111748261769/20502262/LLI-WPS-10.pdf.

Grootaert, C., Van Bastelaer, T., 2002. Instruments of the Social Capital Assessment Tool. World Bank, Washington, DC. Retrieved 18.06.2020, from: https://www.urban-response.org/system/files/content/resource/files/main/annex1.pdf.

Grootaert, C., Narayan, D., Woolcock, M., Nyhan-Jones, V., 2004. Measuring Social Capital: An Integrated Questionnaire. World Bank, Washington, DC. Retrieved 18.06.2020, from: http://documents.worldbank.org/curated/pt/515261468740392133/Measuring-social-capital-an-integrated-questionnaire.

Hanifan, L.J., 1916. The rural school community center. Am. Acad. Pol. Soc. Sci. 67 (1), 130–138.

Joshi, A., Aoki, M., 2014. The role of social capital and public policy in disaster recovery: a case study of Tamil Nadu state, India. Int. J. Disaster Risk Reduct. 7, 100–108.

MIDUVI, 2016. Instrumentos Para la Recuperación HABITACIONAL—Damnificados del Terremoto. Retrieved 18.06.2020, from: http://reliefweb.int/sites/reliefweb.int/files/resources/Plan-construyo-Ecuador-MIDUVI-20160603-MP-18560.pdf.
Minamato, Y., 2010. Social capital and livelihood recovery: post-tsunami Sri Lanka as a case. Disaster Prev. Manag. 19 (5), 548–564.
Nakagawa, Y., Shaw, R., 2004. Social capital: a missing link to disaster recovery. Int. J. Mass Emerg. Disasters 22 (1), 5–34.
OCHA, 2016. Earthquake Ecuador-Situation Report No. 15. The United Nations Office for the Coordination of Humanitarian Affairs. Retrieved 18.06.2020, from: https://reliefweb.int/report/ecuador/unicef-ecuador-humanitarian-situation-report-no15-16-january-2017.
Ostrom, E., 2000. Social capital: a fad or a fundamental concept? In: Dasgupta, P., Serageldpp, I. (Eds.), Social Capital—A Multifaceted Perspective. The World Bank, Washington, DC, pp. 172–214.
Oxfam, 2016. Ecuador Earthquake. Retrieved 10.28.2016, from: https://www.oxfam.org/en/ecuador-earthquake.
Portes, A., 1998. Social capital: its origins and applications in modern sociology. Annu. Rev. Sociol. 24, 1–24.
Putnam, R., 2001. Social capital: measurement and consequences. Can. J. Policy Res. 2 (1), 41–51.
Putnam, R.D., Leonardi, R., Nanetti, R.Y., 1993. Social capital and institutional success. In: Making Democracy Work: Civic Traditions in Modern Italy. Princeton University Press, Princeton, pp. 163–185.
Sanyal, S., Routray, J.K., 2016. Social capital for disaster risk reduction and management with empirical evidences from Sundarbans of India. Int. J. Disaster Risk Reduct. 19, 101–111.
Serageldin, I., Grootaert, C., 1999. Defining social capital: an integrating view. In: Dasgupta, P., Serageldin, I. (Eds.), Social Capital—A Multifaceted Perspective. The World Bank, Washington, DC, pp. 40–58.
SGR, 2014. Agenda Sectorial de Gestión de Riesgos. Secretaría de Gestión de Riesgos, Quito, Ecuador. Retrieved 12.22.2016, from: http://biblioteca.gestionderiesgos.gob.ec/items/show/25.
SGR, 2016a. Informe de situación N°65—16/05/2016—Terremoto 7.8°—Pedernales. Secretaría de Gestión de Riesgos, Quito, Ecuador. Retrieved 18.06.2020, from: http://www.gestionderiesgos.gob.ec/wp-content/uploads/downloads/2016/05/Informe-de-situación-n°65-especial-16-05-20161.pdf.
SGR, 2016b. Resolución No SGR-051-2016. Secretaría de Gestión de Riesgos, Quito, Ecuador. Retrieved 18.06.2020, from: http://www.gestionderiesgos.gob.ec/wp-content/uploads/downloads/2016/05/Resolución-No-SGR-051-2016.pdf.
SNGR, 2016. Informe de gestión 2012. Secretaría Nacional de Gestión de Riesgos, Quito, Ecuador, Retrieved 18.06.2020, from: http://www.gestionderiesgos.gob.ec/wp-content/uploads/downloads/2014/03/Informe_de_Gestion_2012.pdf; 2012.
Son, J., Lin, N., 2008. Social capital and civic action: a network-based approach. Soc. Sci. Res. 37 (1), 330–349.
Telesur, 2016. Reconstrucción de Ecuador tras Terremoto ha sido Impresionante. Retrieved 18.06.2020, from: http://www.telesurtv.net/news/Reconstruccion-de-Ecuador-tras-terremoto-ha-sido-impresionante-20161016-0017.html.
Uphoff, N., 2000. Understanding social capital: learning from the analysis and experience of participation. In: Dasgupta, P., Serageldin, I. (Eds.), Social Capital—A Multifaceted Perspective. The World Bank, Washington, DC, pp. 215–249.
Woolcock, M., 1998. Social capital and economic development: toward a theoretical synthesis and policy framework. Theory Soc. 27 (2), 151–208.

Chapter 12

2017 Coastal El Niño in Peru: An opportunity to analyze the influence of hazard mitigation plans on local resilience

Juan N. Urteaga-Tirado[a], Sandra Santa-Cruz[a], Graciela Fernández de Córdova[b], and Marta Vilela[b]
[a]*GERDIS Research Group, Department of Engineering, Pontifical Catholic University of Peru, Lima, Perú,* [b]*Department of Architecture, Pontifical Catholic University of Peru, Lima, Perú*

1 Introduction

In accordance with the United Nations Office for Disaster Risk Reduction (UNISDR), disaster risk information has been growing exponentially; however, the related economic losses are not yet showing reduction signs, and, in many cases, the post-disaster recoveries have reconstructed equal or greater vulnerabilities (UNISDR, 2015). Considering that both losses and recovery are two important factors in resilience (UNISDR, 2015), the existence of significant deficiencies in the process in which the disaster risk information is translated into a resilience increase may be inferred.

The Hazard Mitigation Plans (HMPs) are instruments using risk information for helping localities understand and assume the factors of such risk and defining strategies and actions for reducing it (Masterson et al., 2014a). In the majority of cases, its documents consist of a first part dedicated to the diagnosis, where risk, vulnerability and hazard estimations are included; as well as a second part where mitigation proposals are shown for being executed in a short, medium or long term. HMPs go through the formulation implementation and update stages.

Generally, studies assessing the HMPs quality of content of various regions in the world have been performed *ex ante* through specific protocols that have been standardizing over time (Lyles and Stevens, 2014). The majority of these studies have been developed in the United States and for multihazards HMPs (cf. Horney et al., 2012, 2016; Janes, 2013; Kang et al., 2010; Lyles et al., 2014; Silapapiphat, 2015), but also in other countries such as South Korea, Canada and Japan and for specific hazards, such as flood or avalanches (Kim and Kakimoto, 2016; Kozel, 2015; Park, 2016; Stevens and Shoubridge, 2015).

Certain studies have analyzed the relation between the HMPs quality and the reduction of direct damages caused by past disasters. For instance, Kang (2009) associated the inclusion level of floods mitigation measures in the urban plans of the Florida State, United States, to the historical economic losses related to the hazard in insured properties, founding that there was no significant correlation. The author indicates that the losses reduction would be mainly influenced by the implementation of plans, land use management and property insurance policies.

The existing relation between HMPs quality and the recovery rapidity, which constitutes one of the two main elements in resilience definition, has not been considered by the aforementioned studies.

Therefore, the objective of this research is to present an *ex post* methodology analyzing the influence of HMPs in (1) the level of direct damages and (2) the recovery rapidity after a disaster in the influence area of an event. For analyzing its applicability, HMPs of various populated areas or localities in Peru affected by the 2017 coastal El Niño heavy rainfalls were assessed.

2 Theoretical framework

2.1 Disaster risk management plans

In accordance with Masterson et al. (2014b), the Disaster Risk Management (DRM) processes are the following: mitigation, preparedness, response and recovery. Mitigation is aimed at actions reducing the risks (e.g., construction of dams, rivers canalization, land use zoning, education campaigns for maintaining and preserving slopes and stream beds). Preparedness

Strengthening Disaster Risk Governance to Manage Disaster Risk. https://doi.org/10.1016/B978-0-12-818750-0.00012-X

focuses on those activities prior to the impact allowing to obtain an acceptable preparation in case of an extreme event (e.g., to raise awareness in the population through education and drills, resource storage, design and installation of early warning systems and monitoring). The response is the activation of plans seen in the preparedness, according to the extreme event reality and starting from the moment when the event is detected until the situation is stabilized after the impact (e.g., affected area protection, search and rescue of persons and medical care provision). Recovery involves the actions taken for rehabilitating and reconstructing the affected properties and for restoring the activities back to normal (e.g., reconstruction of dwellings, highways, bridges and economic system reactivation).

For conducting risk management processes, the local governments develop HMPs, Emergency Operation Plans (mixing the preparedness and response stages) and Recovery Plans. The DRM plans of various localities in the world have been the object of study by various authors. A review of these studies reveals that from the 712 plans evaluated in the last decade, 71% are HMPs, 15% are recovery plans and 14% are Emergency Operation Plans (Table 12.1). The greater attention of the authors toward HMPs is consistent with the guidelines of the international bodies prioritizing risk reduction and prevention strategies over the reactive management (cf. UNISDR, 2015).

2.2 HMPs and their assessment

HMPs have been assessed with the method called content analysis. This method consists of the verification of the presence or absence of specific contents in the documents of plans by means of a form organized into evaluation items (Lyles and Stevens, 2014). This methodology is also used for evaluating climate change, transport, housing and urban plans, etc. The HMPs evaluation items are grouped into categories as those shown in Table 12.2 of the study of Kang et al. (2010) and Peacock et al. (2009). The second column of such table shows the number of evaluation items in each category, and the third column shows the definition of each category. It is observed that the categories of Capabilities and inter-organizational coordination (CIi) and Proposed policies and actions (PAi) comprise 63% of the total evaluation items.

2.3 Resilience and HMPs

In accordance with UNISDR (2016), resilience is the "capacity of a system, community or society exposed to a threat for appropriately and efficiently resisting, absorbing, adapting, transforming and recovering from the effects of such threat." Cimellaro et al. (2016) define resilience as the area under the curve of functionality-performance of a system in the recovery

TABLE 12.1 DRM plan quality evaluation studies in the last 10 years.

DRM process	Studies	Country	# Evaluated plans	Type of hazard
Recovery	Song et al. (2017)	China	16	Earthquake
	Berke et al. (2014)	United States	87	Multiple
Emergency preparedness and response	Meyerson (2012)	United States	3	Multiple
	Saunders et al. (2015)	New Zealand	99	Multiple
Mitigation (HMP)	Kang et al. (2010)	United States	12	Multiple
	Lyles et al. (2014)	United States	175	Multiple
	Horney et al. (2016)	United States	84	Multiple
	Kim and Kakimoto (2016)	United States, Japan and South Korea	3	Flood
	Stevens and Shoubridge (2015)	United States, Canada	20	Multiple
	Janes (2013)	United States	34	Multiple
	Silapapiphat (2015)	United States	76	Multiple
	Kozel (2015)	United States	24	Avalanche
	Park (2016)	United States	49	Flood
	Horney et al. (2012)	United States	30	Multiple

TABLE 12.2 Quality categories of HMPs.

Plan quality category	N° of items	Definition
Vision statement (Vi)	5	Evaluate the extent to which the plan describes the locality, the main hazards to be attended, and the projected physical safety vision.
Planning process (PPi)	10	Evaluate the extent to which the plan describes how was developed and if public participation techniques were used.
Fact base studies (FBi)	21	Evaluate to what extent the hazards to which the locality is exposed, its vulnerability and risk including maps and calculations have been identified. It is considered the main category.
Proposed goals and objectives (GOi)	11	Evaluate to what extent the plan detailed the goals to be achieved, such as the reduction of economic, infrastructure or human losses due to disaster.
Capabilities and inter-organizational coordination (CIi)	29	Evaluates to what extent the plan diagnoses the internal capacities of local governments, and the level of coordination existing with different external agencies to ensure the execution of the proposed measures.
Proposed policies and actions (PAi)	75	Measure the amount and detail of proposed policies and actions that the HMP includes. For example: regulation of land use, structural, educational, economic measures, etc. It is also one of the most important categories.
Implementation and monitoring assurances (IMi)	13	Evaluate to what extent the plan facilitates the implementation of their proposals, as well as the establishment of mechanisms to monitor and update the plan.

(Based in Kang, J.E., Peacock, W.G., Husein, R., 2010. An assessment of coastal zone hazard mitigation plans in Texas. J. Disaster Res. 5(5), 520–528. https://doi.org/10.20965/jdr.2010.p0526 y Peacock, W.G., Kang, J.E., Husein, R., Burns, G.R., Prater, C., Brody, S., Kennedy, T., 2009. An Assessment of Coastal Zone Hazard Mitigation Plans in Texas. Retrieved from: http://hrrc.arch.tamu.edu/_common/documents/09-01R_An_assessment_of_CZ_Haz_Mit_Plans_January_11,_2009.pdf.)

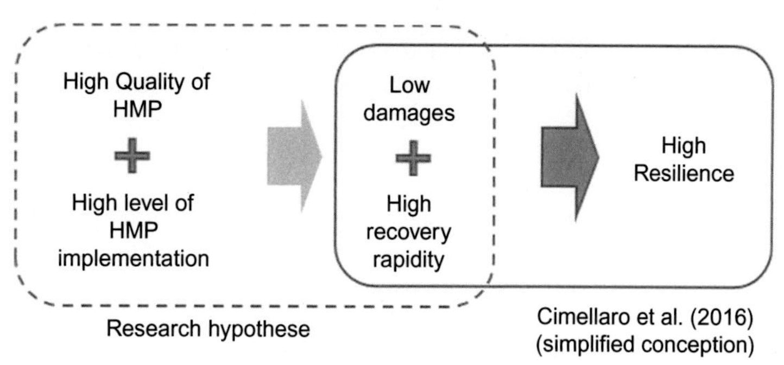

FIG. 12.1 The HMPs and resilience.

time interval after a disaster. Thus, the lower the damage and the greater the recovery rapidity, the greater the resilience. As shown in Fig. 12.1 these resilience elements may be influenced by the quality of existing HMP and its level of implementation. When the disaster event occurs, the current HMP implementation might be interrupted until the emergency attention stage is over; after that, the plan might be updated during or after the recovery term, which will allow the use of the disaster lessons learned in the HMPs update.

3 General proposed methodology

The proposed methodology for analyzing *ex post* the influence of HMPs in the resilience of localities consists of two levels (Fig. 12.2). In the first level, called extensive analysis, a sample of localities affected by a disaster event is selected and the correlations between the variables related to HMPs are analyzed: quality of plans and implementation level; and variables related to resilience: level of damages and recovery rapidity. In the second level called comprehensive analysis, a sub-sample of the localities is collected (e.g., the locality of greater and lower level of damages) for the application of various techniques, mainly qualitative (e.g., detailed interviews, spatial contrast and document review) for better understanding the mechanisms developed among the variables.

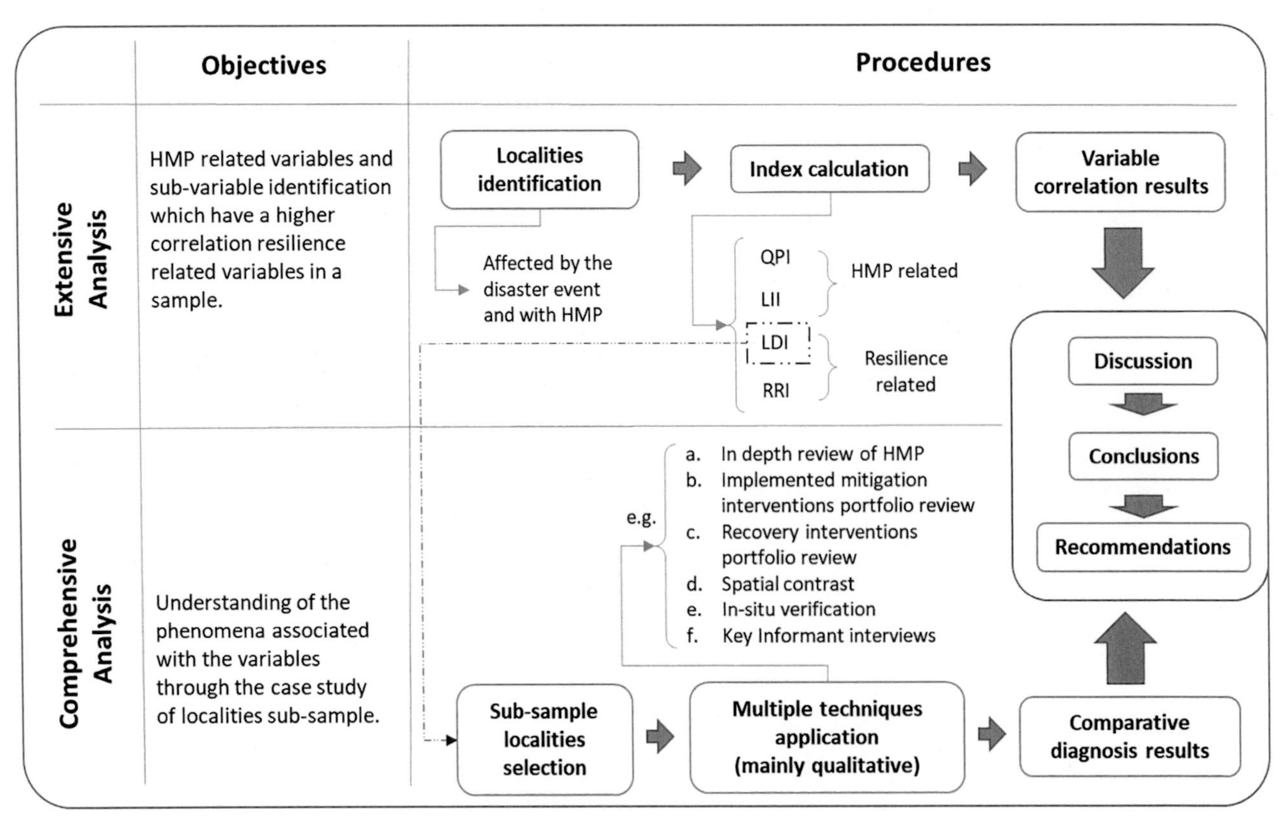

FIG. 12.2 Proposed methodology to analyze the influence of the quality of the HMPs on the resilience of affected localities.

The obtained results of both analysis levels are interpreted jointly for elucidating cause-effect relations. Finally, recommendations as contributions for strategies and actions of hazard mitigation are provided.

3.1 Extensive analysis

In the extensive analysis, the first step is to identify the localities having HMP and that have been considerably affected after a disaster event. The second step is to calculate the indexes of the analysis variables for each locality. Such indexes are properly normalized in order to achieve the comparison among the localities. Such indexes are described below:

(a) *Quality of Plan Index (QPI)*—Numerical and standardized measure of the quality of the plan documents. Therefore, the content analysis protocols as those indicated in Table 12.1 may be used.

(b) *Level of Implementation Index (LII)*—Annual number of executions of mitigation actions from the approval of HMP up until before the event. In case these interventions were not well registered, some other data may be used, for instance, the DRM municipal expenses register. Thus, the following Eq. (12.1) is proposed:

$$LII = \frac{AS}{YI * Pop * LR} \tag{12.1}$$

where AS is the amount spent on mitigation between the year when the HMP was developed until the year before the event, Pop is the locality population, LR corresponds to the locality risk level, and YI is the number of years when AS was measured.

(c) *Level of Damage Index (LDI)*—Normalized rate of direct damages in regard to the locality size (e.g., population and gross domestic product) and to the event severity, Eq. (12.2). This index allows to compare which localities have had a better or worst performance related to the damages as a result of the disaster event. Direct damages may be expressed in terms of victims, dead people, destroyed dwellings, etc.

$$LDI = \frac{Ddam}{Pop * TI} \tag{12.2}$$

where $Ddam$ are the direct damages and TI is the threat intensity.

(d) *Recovery Rapidity Index (RRI)*—Execution percentage of works and activities of recovery in regard to the planned interventions for recovering the locality divided by the recovery time, Eq. (12.3). This index allows to identify, in average, which localities are recovering faster after the disaster event.

$$RRI = \frac{\sum EI_i}{RT * \sum BA_i} \tag{12.3}$$

where EI_i is the expense incurred in the locality in the i-th recovery intervention, BA_i is the i-th intervention amount in the recovery plan portfolio, and RT is the recovery time of each locality. If RRI is estimated before the end of the recovery stage, such time would be the same for all the localities; therefore, in that case, such factor may be disregarded.

In the third step, the statistical analysis is conducted by identifying those variables related to HMPs (QPI and LII) having a better correlation with resilience variables (LDI and RRI).

3.2 Comprehensive analysis

The comprehensive analysis level begins with the selection of a sub-sample of the localities selected in the extensive analysis. Secondly, various analyses allowing to deeper understand if the plans quality has allowed to develop better mitigation actions, and if such actions have achieved effectively to be translated into damages reduction and recovery rapidity increase, are developed. The analyses to be used may be those described herein below:

(a) *HMP In-Depth Review*—A deeper review is conducted in comparison with the review carried out in the content analysis developed in the extensive level. A summarized description of the HMP main characteristics is obtained (e.g., HMP organization, main findings in the diagnosis and mitigation proposals, identification of which interventions proposed by the HMP would be key to resilience).
(b) *Review of the mitigation intervention portfolio implemented before the event*—The objective of this review is to determine if the mitigation interventions that have been implemented after the HMP approval are related to the plan.
(c) *Recovery Intervention Portfolio Review*—The recovery interventions before the event that have been planned and executed in the affected localities are reviewed for determining if they are related to the HMPs evaluated in the extensive level.
(d) *Spatial Contrast*—The coherence between the different available spatial information sources is verified (e.g., cadastres of affected dwellings, land use zoning maps, risk maps, hazards, vulnerability, landforms, etc.). Likewise, to spatialize the key information obtained in the previous analyses is also important. All this for better interpreting the land complexity, as well as to identify key areas for performing the inspection in the field and interviews to key informants.
(e) *In situ Verification*—It consists of the development of inspections of both localities and their identified key areas allowing to know and verify the information analyzed in the spatial contrast, the review of the recovery and mitigation works portfolio and the HMP in-depth review.
(f) *Key Informants' Interviews*—It consists of the development of interviews, focus groups, workshops or work groups with local actors (i.e., authorities and officers of the local governments, local technical specialists, representatives of production activities, leaders and neighbors of the identified key areas in the spatial contrast) for validating the obtained data and to receive first-hand information. To consider the officers responsible for the health, education, social management, urban planning, budgets, works and risk management is important.

Finally, a comparative analysis between the findings in the localities identifying the good and bad practices or factors that have influenced the effects and the recovery rapidity of both localities was carried out.

4 Application to the 2017 coastal El Niño event in Peru

This event of heavy rainfalls that occurred between the months of December 2016 and May 2017 is considered to be the third most intense "El Niño Phenomenon" in the last 100 years in Peru (ENFEN, 2017). Such phenomenon left 169 dead people, 283.137 victims, 1.644.879 affected people and economic losses of 1.6% of Peru's Gross Domestic Product (GDP) (INDECI, 2018; PCM, 2017b). Fig. 12.3 shows the precipitation anomaly percentage values as measure of the event severity (cf. SENAMHI, 2017a) in the affected region. As observed, the biggest threat was focused on the coastal region of the country with anomaly values up to 2000% except for Tumbes department, where the values were moderate and almost null in the Ecuadorian border. As a result of the coastal El Niño event magnitude, the government elaborated a national reconstruction plan (PIRCC), with an allocated budget of 7.8 billion dollars (3.7% of GDP) from which 21% was allocated to disaster prevention works (PCM, 2017a).

The most affected area by this event was the North coast of the country covering five regions. Many of the affected localities had HMPs elaborated as part of a nationwide initiative involving 177 localities of Peru. The initiative was called "Sustainable Cities Programme" (PCS), and it was developed between 1998 and 2015 with the help of the United Nations

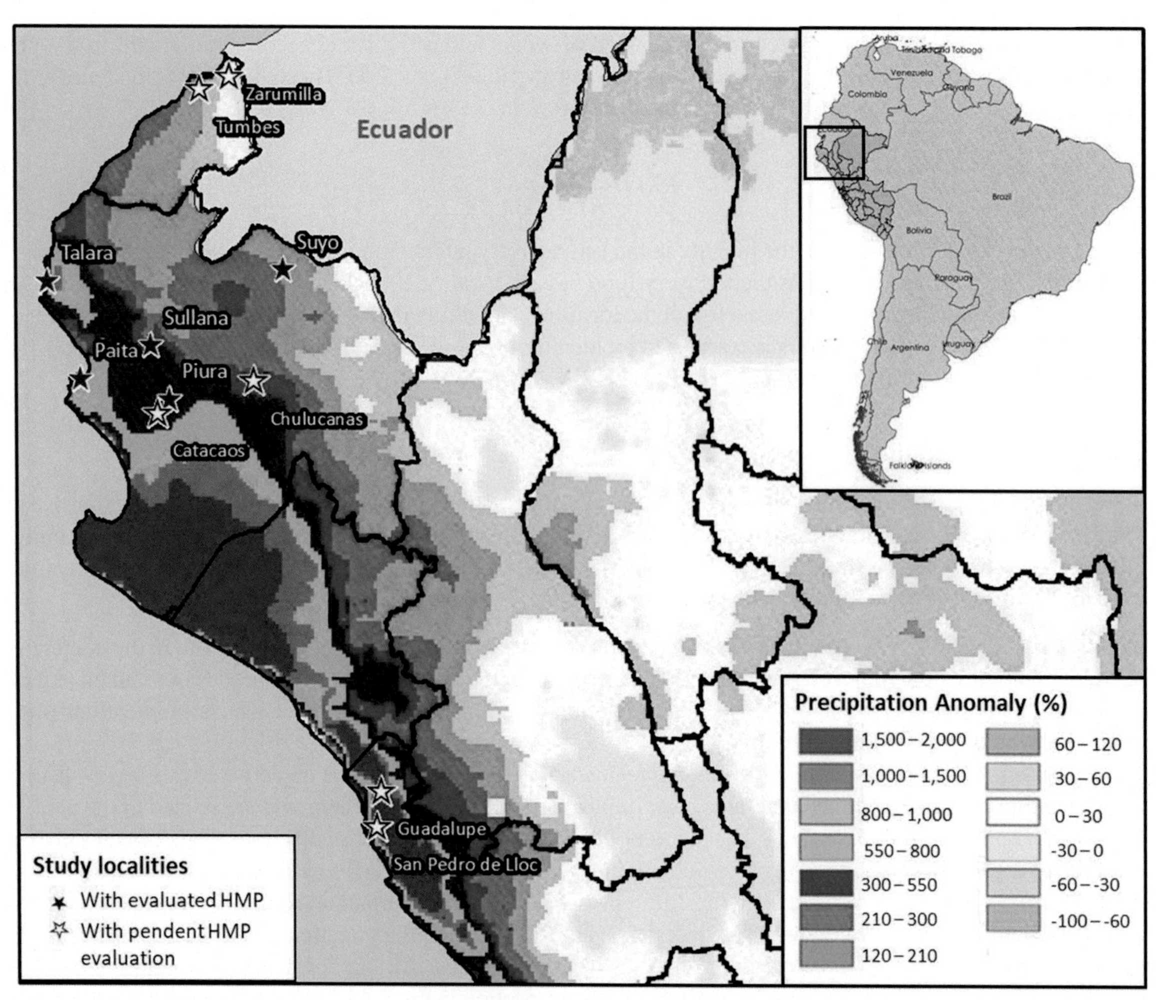

FIG. 12.3 Precipitation level in terms of precipitation anomaly percentage from January to March 2017 and localities of the study. *(Based on SENAMHI, 2017b. Monitoreo de Lluvias en el Norte Enero—Febrero—marzo 2017. Servicio Nacional de Meteorología e Hidrología del Perú (SENAMHI). Retrieved from: http://www.senamhi.gob.pe/load/file/02231SENA-10.pdf.)*

Development Programme (UNDP) and the Organization of American States (OAS) (INDECI, 2016a). In the last decade, PCS was assessed two times, once domestically by *Instituto Nacional de Defensa Civil* (National Institute of Civil Defence, INDECI) and UNDP (2013) and a second time externally by Lavell (2016). The first assessment highlighted the importance of public participation and the capacities strengthening and monitoring and implementation of the mitigation actions were identified as the main challenges. The second assessment recognized PCS as a great national information contribution; however, the fact of not prioritizing the institutionalization of such initiative was criticized, and weaknesses in the appropriation and implementation of the plans by the local governments were identified (cf. Lavell, 2016). Although these assessments provide valuable recommendations for improving and implementing HMPs, the studies used a qualitative approach and an expert judgment, and reviewed a small sample of all the existing plans and the statistical data of emergencies or investments in DRM was not considered.

4.1 First level—Extensive analysis

Localities affected by the coastal El Niño event that had a maximum 10-year-old HMPs before the event occurred and that were part of coastal departments of the country Northern region were selected for guaranteeing similar regulations and geographical conditions. For being considered affected, localities must have had both number of victims and precipitation anomaly percentage above zero. Selected localities may be seen in Table 12.3 and Fig. 12.3.

Variables are calculated according to the following:

(a) *Quality of Plan Index (QPI)*—A HMPs content analysis protocol was applied based on Kang et al. (2010), since it had a more detailed item codebook of the reviewed literature (cf. Peacock et al., 2009). Adaptations conducted include: translation, summarized application for multiple risks and adaptation to the local regulations.

TABLE 12.3 Selected localities for extensive analysis.

N°	Locality	Department	Year of the HMP	Population 2015	HMP Evaluation
1	Zarumilla	Tumbes	2008	22,500	Pendent
2	Tumbes	Tumbes	2011	88,360	Pendent
3	**Suyo**	**Piura**	**2008**	**1500**	**Finished**
4	**Paita**	**Piura**	**2011**	**57,437**	**Finished**
5	**Talara**	**Piura**	**2010**	**135,000**	**Finished**
6	**Sullana**	**Piura**	**2010**	**180,000**	**Finished**
7	**Piura**	**Piura**	**2011**	**450,363**	**Finished**
8	Catacaos	Piura	2010	50,419	Pendent
9	Chulucanas	Piura	2010	55,183	Pendent
10	San Pedro de Lloc	La Libertad	2010	12,171	Pendent
11	Guadalupe	La Libertad	2009	20,703	Pendent

Note: Localities in bold means that the extensive analysis is completed.

(Based on INDECI, 2016a. Compendio Estadístico del INDECI 2016, en la Preparación, Respuesta y Rehabilitación ante Emergencias y Desastres. Lima: Instituto Nacional de Defensa Civil (INDECI). Retrieved from: https://www.indeci.gob.pe/listado.php?itemC=NDY=&item=OTUz.)

Each item may be rated with values 0, 1 and 2 corresponding to the level in which the content is complied with (e.g., 0 = without mention, 1 = mention without details, 2 = detailed mention). After that, the addition of the obtained scores of each item by category (Table 12.2) is performed, dividing it by the possible total score to be obtained in such category; thus, a percentage rating is obtained. Then, those scores are averaged as shown in Eq. (12.4):

$$QPI = \frac{Vi + PPi + FBi + GOi + CCi + PAi + IMi}{7} \tag{12.4}$$

(b) *Level of Implementation Index (LII)*—This index was calculated with Eq. (12.1). *AS* was calculated as the amount spent in DRM between the year when the HMP was developed and year 2016 obtained from the Economic Transparency Site of *Ministerio de Economía y Finanzas* (Ministry of Economy and Finance) (MEF) (n.d.-b) in Budgetary Category No. 0068 (cf. PCM, 2018). *Pop* variable was calculated through the addition of population of districts conforming the locality in accordance with *Instituto Nacional de Estadística e Informática* (National Institute of Statistics and Informatics) (INEI) (n.d.); *LR* was calculated as the annual average number of victims between 2003 and 2016 obtained by INDECI.

(c) *Level of Damage Index (LDI)*—The victims normalized number was used. A victim is a person whose dwelling was declared unhabitable or destroyed (INDECI, 2016b). This index is calculated according to Eq. (12.2). *Ddam* value was calculated as the number of victims related to the event according to INDECI, *TI* was calculated as the precipitation anomaly percentage in accordance with *Servicio Nacional de Meteorología e Hidrología del Perú* (National Service of Meteorology and Hydrology of Peru) (SENAMHI) (2017b).

(d) *Recovery Rapidity Index (RRI)*—A percentage normalization of budgetary execution corresponding to the reconstruction plan was used in accordance with Eq. (12.3). *EI* is the expense incurred corresponding to the PIRCC in each district of a locality according to MEF (n.d.-a) and *BA* is the calculated amount in PIRCC portfolio for locality districts in accordance with Authority for Reconstruction with Changes (RCC) (2017). The majority of PIRCC amounts were considered, except for those that did not have a clear allocation for a locality. The measurement was carried out before ending the reconstruction stage (TR = 22 months); therefore, TR value is constant for all localities.

4.2 Second level—Comprehensive analysis

Two cities were selected to be analyzed, one with the greater and the other with the lower LDI.

(a) *HMPs In-Depth Review*—Assessed plans were obtained from INDECI and *Centro Nacional de Estimación, Prevención y Reducción del Riesgo de Desastres* (National Centre for Disaster Risk Estimation, Prevention and Reduction) (CENEPRED) public websites. Furthermore, they were compared to plans found when performing interviews to local officers.

(b) *Review of the mitigation intervention portfolio implemented before the event*—The intervention list corresponding to DRM registered in the MEF transparency site (n.d.-b) was analyzed, in Budgetary Category No. 0068 until year 2016 for districts conforming the locality.

(c) *Recovery Intervention Portfolio Review*—The reconstruction intervention list corresponding to PIRCC (cf. RCC, 2017) was reviewed for districts conforming the locality by mainly observing their description, typology and budgetary magnitude.

(d) *Spatial Contrast*—Two layers were compared: spatial distribution of dwellings affected by the coastal El Niño according to *Comisión de Formalización de la Propiedad Informal* (Commission for the Official Registration of Informal Property) (COFOPRI) (n.d.-a), current land use zoning plan of each locality, interventions location prioritized by HMP, HMP hydrologic hazards map (since there was not an hydrological risk map), stream beds location, risks summary map and alluvial plains of past events. With such information, three key areas for in situ verification and interviews with leaders and neighbors were determined:

- Area 1—It combines a high risk with a large number of negative effects. It may represent that since HMP was approved, no action has been executed for reducing the estimated risk.
- Area 2—It combines a low risk with a significant quantity of negative effects. It may be related to a deficient risk estimation in HMP or that vulnerability has increased over time.
- Area 3—Optional, but indicating justification (e.g., HMP key intervention location, conflicts with land use zoning, significant landforms, etc.).

(e) *In situ Verification*—Inspections to selected key areas and localities were carried out. Annotations, video and photographic recordings were developed.

(f) *Key Informants Interviews*—Due to limited time availability, only interviews to DRM and urban management officers were possible, as well as interviews to leaders and certain key areas neighbors.

5 Results and discussion

5.1 Extensive analysis

Table 12.4 shows QPI obtained values (with a mean of 37% and a standard deviation of 5%). The average value is similar to the value of city HMPs assessed by Kang et al. (2010), that was 35%. QPI values vary among localities from 29% in Talara to 43% in Suyo; and among categories, from 23% in CIi to 62% in Vi, on average for all localities. Talara's low rating is mainly caused by the poor performance in CIi and GOi categories. Suyo has an outstanding rating mainly because of Vi, PPi and GOi categories, since its HMP was elaborated with social participation, coordination between officers (even from a border city of Equator) and was widely spread among the population. All HMPs, except for Suyo, obtained a low rating in CIi category, being more than 10 points below the cities of the comparison study.

Table 12.5 shows LII, LDI and RRI values in selected localities. Two localities (Suyo and San Pedro de Lloc) have null LII since they do not register expenses in DRM. However, Suyo had 92 victims per year in the 2003–16 term, compared to

TABLE 12.4 Plan quality results and comparison with Kang et al. (2010) cities.

N°	Plan quality category	Suyo	Paita	Talara	Sullana	Piura	Mean	Kang et al. (2010)
1	Vision statement (Vi)	70%	60%	40%	70%	70%	62%	37%
2	Planning process (PPi)	45%	20%	20%	15%	25%	25%	45%
3	Fact base studies (FBi)	43%	43%	50%	62%	45%	49%	21%
4	Proposed goals and objectives (GOi)	45%	32%	18%	36%	36%	34%	27%
5	Capabilities and inter-organizational coordination (CIi)	33%	21%	17%	21%	24%	23%	36%
6	Proposed policies and actions (PAi)	40%	37%	38%	39%	35%	38%	28%
7	Implementation and monitoring assurances (IMi)	27%	31%	19%	23%	31%	26%	49%
	Quality of plan (QPI)	**43%**	**35%**	**29%**	**38%**	**38%**	**37%**	**35%**

TABLE 12.5 Variables indexes calculations.

Data	Suyo	Paita	Talara	Sullana	Piura	Catacaos	Chulucanas	Zarumilla	Tumbes	San Pedro de Lloc	Guadalupe
Districts involved	Suyo	Paita	Pariñas	Sullana, Bellavista	Piura, Castilla, 26 de Octubre	Catacaos	Chulucanas	Zarumilla	Tumbes	San Pedro de Lloc	Guadalupe
2017 Population (*Pop*)[a]	12,377	95,980	90,592	217,867	454,057	74,002	76,238	22,895	114,300	16,815	44,496
2017 Event victims (*Ddam*)[b]	50	4360	3000	427	26,720	28,707	3090	85	644	5398	254
A. precipitation % (*TI*)[c]	135	775	1025	255	425	425	272	15	86	1650	1750
LDI ($\times 10^{-4}$)	**0.3**	**0.59**	**0.32**	**0.08**	**1.38**	**9.13**	**1.49**	**2.48**	**0.66**	**1.95**	**0.03**
USD invested 2012–16 (*AS*)[d]	0	3,956,383	3,247,820	1,171,608	1,383,183	26,860	98,436	104,711	3,181,071	0	46,315
Annual average of victims (*LR/YI*)[e]	91.6	64.0	36.6	232.4	539.4	66.6	88.8	13.3	60.4	0	0.1
LII	0	12,364	17,762	1008	513	81	222	1576	10,528	0	64,841
Reconstruction Budget (*BA*)[f] ($)	14,092,730	41,454,450	16,466,980	88,563,129	149,022,380	76,743,252	89,957,359	3,035,488	33,022,298	18,471,979	7,449,204
Executed reconstruction budget (*EA*)[g] ($)	1,055,382	4,301,678	61,195	2,574,977	8,382,990	3,353,285	1,384,640	0	4,871,710	0	0
RRI	7.5%	10.4%	0.4%	2.9%	5.6%	4.4%	1.5%	0.0%	14.8%	0.0%	0.0%

YI = 5 years.
[a] *INEI (n.d.).*
[b] *INDECI.*
[c] *SENAMHI (2017b).*
[d] *MEF (n.d.-b).*
[e] *INDECI.*
[f] *RCC (2017).*
[g] *MEF (n.d.-a).*

San Pedro de Lloc registering victims only in the 2017 coastal El Niño event. In this last case, San Pedro de Lloc local government was likely to have problems for registering effects of past emergencies. Zarumilla and Guadalupe localities have the lowest reconstruction budgets and they have not yet been executed. San Pedro de Lloc has not yet elaborated the budget either, although it has a larger budget. It is important to consider that until 2018, only 8.5% of PIRCC budget (cf. MEF, n.d.-a) was executed in all the Peruvian territory; thus, localities execution percentages were also low. It is also observed that Paita and Tumbes, localities which have applied reconstruction execution the most, have had a good execution in LII.

With the values of Table 12.5, correlations between QPI, LII, LDI and RRI were calculated. Although the quality of plans was expected to have an influence in direct damages, results showed there is no correlation between QPI and LDI. However, a good correlation of LDI with quality categories was found, PAi ($\rho=0.89$) measuring the quantity and detail of policies and actions proposed in HMPs (Fig. 12.4A), and FBi ($\rho=0.57$) measuring baseline studies quality.

HMPs quality was expected to have an influence on recovery rapidity, since it is likely that the information provided by HMPs (e.g., construction safety areas, key mitigation interventions and key contact institutions) would allow the local government to have a better understanding on reconstruction process. However, QPI and RRI correlation was medium ($\rho=0.49$). In a disaggregated analysis, IMi quality category was observed to have a better correlation with RRI ($\rho=0.87$) (Fig. 12.4B). This result means that HMPs contents, among others, related to the identification of those responsible for implementing interventions, its monitoring, possible financing funds and intervention costs estimation would be the most relevant aspects for increasing recovery rapidity.

An exponential relation between LII and LDI with $\rho=0.64$ has been identified, showing that localities that spent more in DRM in the last 5 years presented less damage that those that spent less (Fig. 12.4C). LII index may be improved if aspects related to expense quality are considered.

Similarly, LII and RRI correlation of the 11 localities showed a very low correlation ($\rho=0.17$). However, when removing the five localities with lower reconstruction budget, correlation increases to 0.88 (Fig. 12.4D), suggesting that LII influence in RRI is greater in localities with considerable reconstruction budgets.

The extensive analysis results indicate that QPI calculation with Eq. (12.4) (i.e., simple average of quality category values) may be improved by giving more significance to PAi and FBi categories. The possibility of increasing categories such as "Organization and Presentation" and "Legal Requirements Compliance" may also be explored, which was used by Guyadeen (2017) in the urban development plan assessment. The localities sample extension for having a larger variety of cases is still to be conducted, for instance, standard deviation of QPI above 5%.

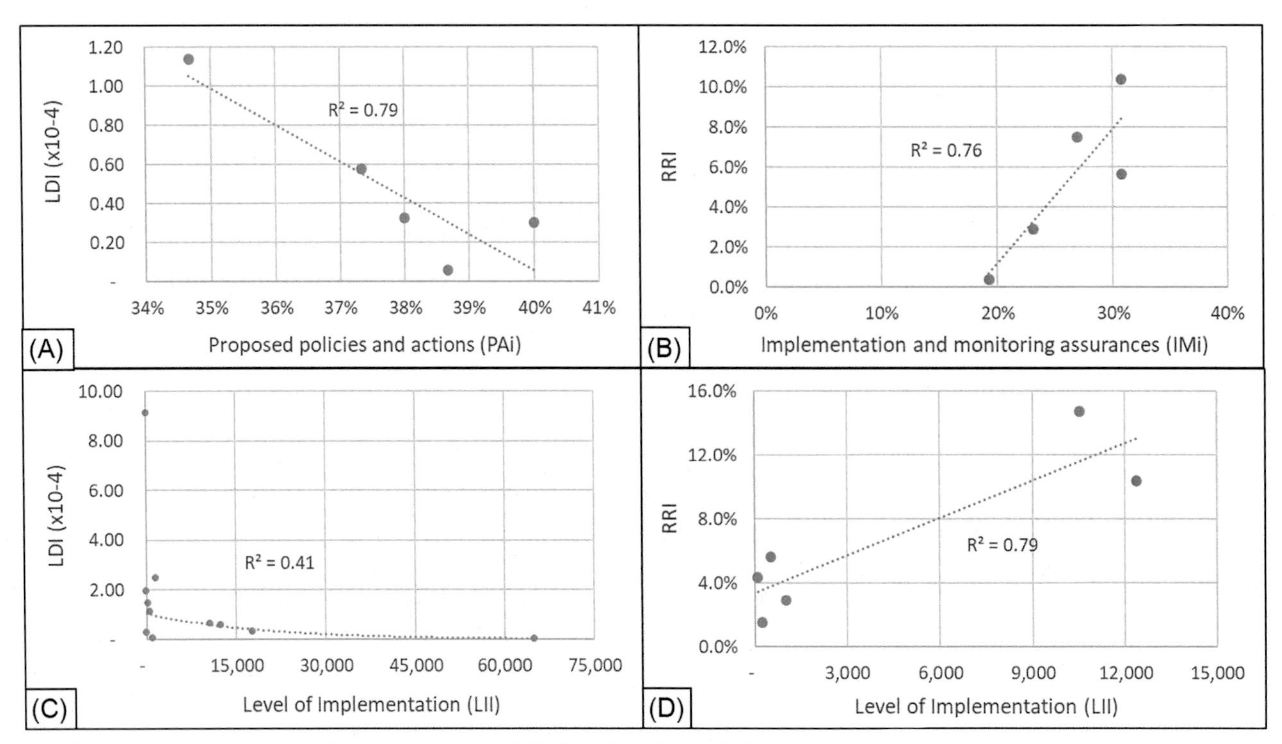

FIG. 12.4 Main correlations: (A) LDI vs PAi. (B) RRI vs IMi. (C) LDI vs LII, (D) RRI vs LII.

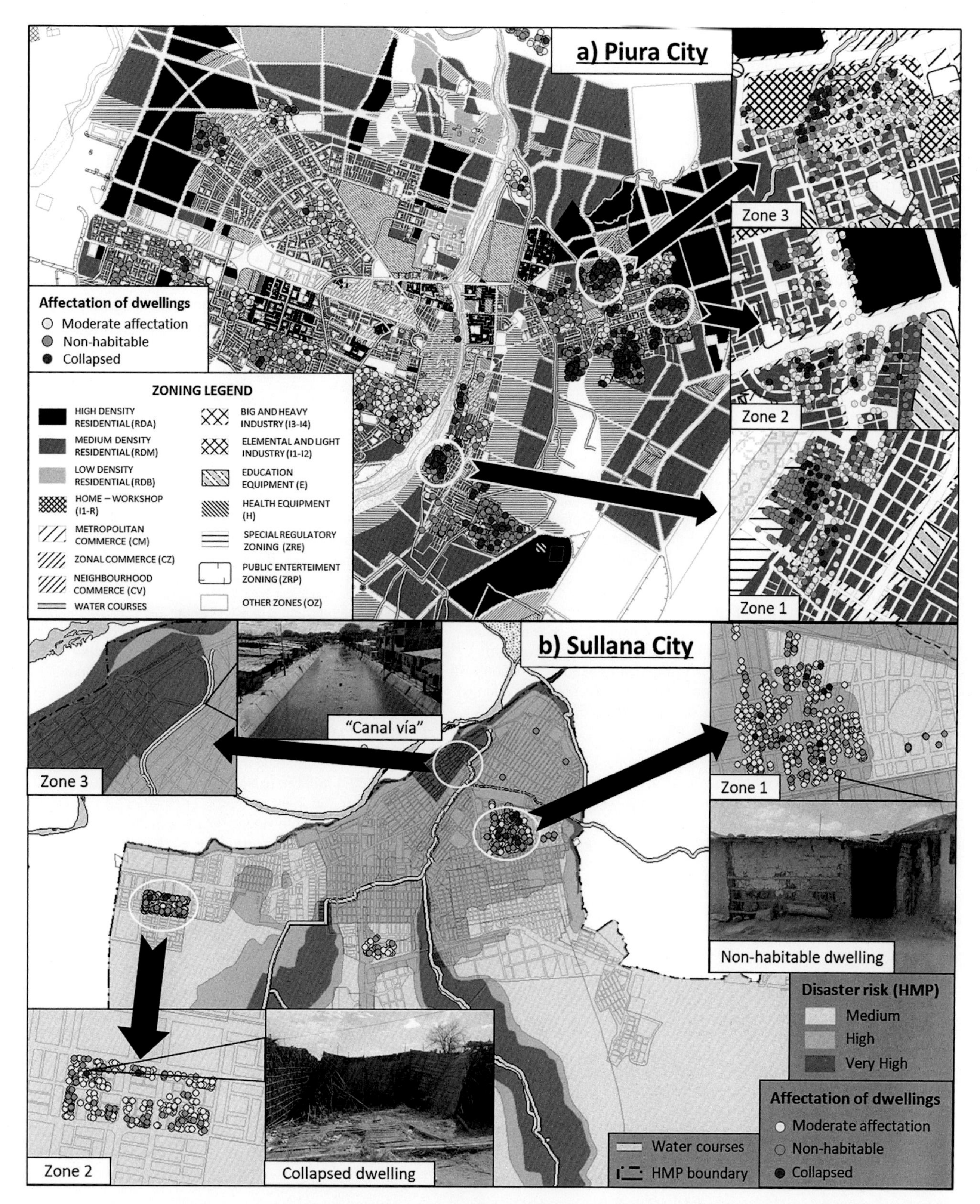

FIG. 12.5 Spatial contrast results. (A) City of Piura. Spatial contrast between the land use zoning map and the distribution of affected dwellings. Data: Provincial Municipality of Piura and COFOPRI (n.d.-a). (B) City of Sullana. Spatial contrast between distribution of affected dwellings and the risk map of the HMP. Data: INDECI, COFOPRI (n.d.-a). Photographs: COFOPRI (n.d.-b) and UDEP (n.d.).

5.2 Comprehensive analysis

Piura and Sullana cities were analyzed, since they were the localities with higher and lower LDI, respectively. The main results of this analysis are contained in Table 12.6. In HMPs in depth review, significant advantages in Sullana related to the quality of plan that were not considered within the used content analysis protocol were found. Although both obtained a QPI rating of 38%, certain internal ratings by category involved differences, highlighting that Sullana FBi is remarkably higher. Among these differences giving superiority to Sullana, we may find: the fact of registering the risk, vulnerability and hazard areas assessment process, as well as budgets estimation of proposed interventions. The following additional advantages that the content analysis protocol did not consider were found: the fact of having been elaborated in one sole stage, having a better content structure and having less insignificant information in its documents.

In the mitigation interventions review implemented before the event, Sullana presented a better performance again. In general, the amount spent in DRM was similar to Piura's amount, although Sullana has half the population of Piura. When comparing its mitigation investment, such aspect is even greater, emphasizing the existence of a structural mitigation work carried out as the one of larger amount. When reviewing the reconstruction interventions portfolio, expenses magnitudes are

TABLE 12.6 Comprehensive analysis comparative results.

Analysis	Piura	Sullana
In-depth review of HMPs	Its HMP was developed in two phases, the first one in 2009 in which was carried out the hazard evaluation and proposition of some interventions; and the second in 2011 that added the vulnerability and risk estimation, and additional intervention proposals that include land use zoning recommendations.	Its HMP was carried out in one phase, with a much better edition of the document than that of Piura (e.g., explanation of the assessment process of the danger, vulnerability and risk zones, less irrelevant content, better structuring). The project files have estimated budgets.
Review of mitigation interventions implemented before the event	The implementation of DRM in this period amounted to USD 1.4 million, of which 22% was executed in mitigation interventions, mainly oriented to building security inspections (USD 0.3 million). The non-mitigation interventions were mainly oriented to emergency preparedness and response (USD 1 million).	The implementation of DRM in this period amounted to USD 1.2 million, of which 44% was executed in mitigation interventions, mainly oriented to the construction of a rainwater drain (USD 0.5 million) and the building security inspections (USD 0.03 million). Non-mitigation interventions were also geared to emergencies (USD 0.6 million).
Review of the portfolio of reconstruction interventions	The 191 corresponding interventions amount to USD 149 million, having executed 5.6% to 2018. A significant amount of the budget is the reconstruction of approximately 2000 homes (USD 14 million) and the improvement of the sewerage system (USD 33 million), the rest of the interventions correspond mainly to infrastructure reconstruction.	The 73 corresponding interventions amount to USD 88.4 million, having executed 2.9% to 2018. The most important amount is oriented to the improvement of sewerage (USD 12 million) and the construction of a hospital (USD 8 million). A reconstruction of 221 houses (USD 1.4 million) is planned and the rest of the interventions are mainly aimed at infrastructure rehabilitation.
Spatial contrast	Negative effects are widespread, with more than 15 zones. The most affected dwellings are found mainly in the eastern zones (district of Castilla), which is why the key areas were located there (Fig. 12.5A). Area 3 chosen corresponds to an area in which the HMP proposed the implementation of a river defense project, which its not implementation implied significant damages. Additionally, it is observed that the current land use zoning did not consider the construction restrictions corresponding to high and very high hazard zones. 701 homes, between uninhabitable and collapsed, were in High or Very High Risk areas.	The affectations are minor and are concentrated in basically 3 zones. Area 3 chosen corresponds to an important mitigation work that exists in the city, called *Canal-vía*, which is a road that only works in the rainy season as a channel (Fig 12.5B). Greater compatibilities were observed between the high and very high hazard zones, with the restrictions of land use zoning. 61 homes, between uninhabitable and collapsed, were in areas of High Risk (in very high risk area it was not found).
In situ verification	It was observed that the river is practically at the same height as the urban environment of its left bank (district of Castilla). That is why many affectations are due to the overflow of this river. Also due to flooded basins and the activation of streams.	It was observed that the height of the river is approximately 15 m below the urban environment. So the effects have been due mainly to the flooded basins and the vulnerability of roofs.

TABLE 12.6 Comprehensive analysis comparative results—cont'd

Analysis	Piura	Sullana
Key informants interviews	The most used HMP documents are those of the first phase (those of the second very little). There was a low organization of the DRM in the municipality of Castilla (e.g., the official in charge of DRM was not dedicated exclusively and it was indicated that in the last 5 years this office had eight officials in charge).	The plan is usually used in its entirety. A better organization was observed than in the Municipality of Castilla. The official in charge of the DRM had exclusive dedication and in the last 5 years there were only two officials in charge. The official founded was the same one who was in charge when the HMP was formulated.

observed to be proportional to their population, although the execution percentage is higher in Piura, probably because of the visibility of its considerable effect or the priority that has been given to its attention. Both coincide in allocating a significant amount for improving sewerage; however, Sullana has more rehabilitation than reconstruction works, unlike Piura.

In the spatial information contrast, damages to dwellings were observed to be more extensive in Piura than in Sullana. Also, a greater compatibility was observed in Sullana between the hazards map and the land use zoning. This coincides with the fact that the number of collapsed and uninhabitable dwellings in a high-risk or very high-risk area, in accordance with HMP, is 10 times larger in Piura than in Sullana.

In the in-situ verification, significant geomorphological differences that may have contributed to a greater impact in Piura were found, such as its greater difference in height between the river and its urban surroundings, and larger presence of flooded basins. These characteristics may better standardize LDI variable.

Interviews to key informants allowed to closely recognize the current use of HMPs by officers. Again, Sullana had a better performance for using its integrity plan (not just a part of it as in Piura), and also due to the full-time attention of officers responsible for DRM and their low turnover in the last 5 years. When collecting information in interviews, the HMPs managed version was observed to be a little more detailed than the one shown on the website.

6 Limitations and lessons learned

It is important to observe that the proposed methodology analyses *ex post* a disaster event in a region due to a specific hazard (in this case, heavy rains) and compares them with current HMPs, although, in general, HMPs are oriented to understand various hazards; results obtained with this methodology may not be extrapolated to other hazards threatening the region (e.g., earthquakes or volcanic eruptions).

A limitation of this particular study is that the plans assessment was performed only by one encoder, unlike other studies where two or even three encoders were used for increasing reliability (cf. Kang et al., 2010; Horney et al., 2016). Moreover, time was short and the interviews with budgetary managing directors and those responsible for the reconstruction remained pending. It would be more efficient that, to the possible extent, the analysis comprehensive level is developed almost simultaneously with the extensive level; thus, more time would be available for performing, at least, two in situ verifications in the localities. In the first one, general land surveying and contact with authorities would be conducted and; in the second one, analysis attention would be focused on key areas.

7 Conclusions

The proposed *ex post* methodology allows to analyze Hazard Mitigation Plans influence on the resilience of localities affected by a disaster event, by means of integrating quantitative and qualitative research methods. The case study showed the proposed method applicability, which may be applied in other regions with similar conditions.

The analysis of damages and reconstruction process after the 2017 Coastal El Niño Phenomenon showed that inclusion in action plans, such as land use regulation, economic, educational and structural reinforcements in HMPs positively influence on the reduction of damages; and that the identification of those responsible for these processes, verification of possible financing funds and intervention costs estimation positively influence on the reconstruction rapidity. Aspects as the general locality descriptions, planned security vision, risk, vulnerability and hazard estimations, have an intermediate influence. Other aspects, such as: explication of proposed objectives, explication of plan formulation process, or explication of the functional capacities of DRM offices found, have less influence.

DRM interventions implementation before the event are confirmed to be related to a damage reduction, as well as to the reconstruction rapidity increase. Likewise, in the comprehensive analysis, the full-time attention of officers responsible for DRM, as well as a low turnover in their positions were also considered as factors influencing resilience.

In future studies, the increase in the number of events and case studies analyzed for reaching conclusions for other phenomena such as earthquakes or hurricanes is recommended.

Acknowledgment

This paper was partially funded by the Peruvian government in the framework of Agreement 232-2015-FONDECYT.

References

Berke, P., Cooper, J., Aminto, M., Grabich, S., Horney, J., 2014. Adaptive planning for disaster recovery and resiliency: an evaluation of 87 local recovery plans in eight states. J. Am. Plan. Assoc. 80 (4), 310–323. https://doi.org/10.1080/01944363.2014.976585.

Cimellaro, G.P., Renschler, C., Reinhorn, A.M., Arendt, L., 2016. PEOPLES: a framework for evaluating resilience. J. Struct. Eng. 142 (10). https://doi.org/10.1061/(ASCE)ST.1943-541X.0001514.

COFOPRI, n.d.-a. Catastro de Daños Fenómeno del Niño Costero. Organismo de Formalización de la Propiedad Informal (COFOPRI). Retrieved July 1, 2019, from: http://geoportal.cofopri.gob.pe/cofopri/rest/services/CDC/FNC/MapServer.

COFOPRI, n.d.-b. Geoportal Geo Llaqta. Plataforma única de catastro multipropósito. Layer: Catastro de daños. Organismo de Formalización de la Propiedad Informal (COFOPRI). Retrieved July 1, 2019, from: https://catastro.cofopri.gob.pe/geollaqta/.

ENFEN, 2017. Informe técnico extraordinario N°001–2017/ENFEN EL NIÑO COSTERO 2017. Comité Multisectorial Encargado del Estudio Nacional del Fenómeno El Niño (ENFEN). Retrieved from: http://www.imarpe.pe/imarpe/archivos/informes/imarpe_inftco_informe__tecnico_extraordinario_001_2017.pdf.

Guyadeen, D., 2017. Evaluation in Planning: An Investigation into Plan Quality and its Application to Official Plans in the Ontario-Greater Golden Horseshoe (GGH) Region (PhD thesis). University of Waterloo. Retrieved from: http://hdl.handle.net/10012/11832.

Horney, J., Naimi, A.I., Lyles, W., Simon, M., Salvesen, D., Berke, P., 2012. Assessing the relationship between hazard mitigation plan quality and rural status in a cohort of 57 counties from 3 states in the Southeastern U.S. Challenges 3 (2), 183–193. https://doi.org/10.3390/challe3020183.

Horney, J., Nguyen, M., Salvesen, D., Dwyer, C., Cooper, J., Berke, P., 2016. Assessing the quality of rural hazard mitigation plans in the southeastern United States. J. Plan. Educ. Res. 37 (1), 56–65. https://doi.org/10.1177/0739456X16628605.

INDECI, 2016a. Compendio Estadístico del INDECI 2016, en la Preparación, Respuesta y Rehabilitación ante Emergencias y Desastres. Instituto Nacional de Defensa Civil (INDECI), Lima. Retrieved from: https://www.indeci.gob.pe/listado.php?itemC=NDY=&item=OTUz.

INDECI, 2016b. Manual de evaluación de daños y análisis de necesidades EDAN—PERÚ. Instituto Nacional de Defensa Civil (INDECI). Retrieved from: http://bvpad.indeci.gob.pe/doc/pdf/esp/doc2662/doc2662-contenido.pdf.

INDECI, 2018. Resumen ejecutivo histórico al 100% de la temporada de lluvias 2016–2017. Instituto Nacional de Defensa Civil (INDECI). Retrieved from: https://www.indeci.gob.pe/objetos/alerta/MjY2MA==/20180903133549.pdf.

INDECI, & PNUD, 2013. Programa Ciudades Sostenibles "Lecciones Aprendidas y Sistematización de Buenas Prácticas" Una Contribución a la Reducción de Desastres en el Perú. Instituto Nacional de Defensa Civil (INDECI), Programa de las Naciones Unidas para el Desarrollo (PNUD), Lima. Retrieved from: http://www.pe.undp.org/content/dam/peru/docs/Prevención y recuperación de crisis/Nuevas publicaciones/pe.CiudadesSostenibles.pdf.

INEI, n.d.. Sistema de Información Regional Para la Toma de Decisiones. Instituto Nacional de Estadística e Informática (INEI). Retrieved July 28, 2018, from: http://webinei.inei.gob.pe:8080/SIRTOD/inicio.html#.

Janes, E.H., 2013. Evaluating Local Hazard Mitigation Planning: Quality, Public Participation, and Social Vulnerability (PhD thesis). University of Colorado. Retrieved from: http://digital.auraria.edu/content/AA/00/00/01/34/00001/AA00000134_00001.pdf.

Kang, J.E., 2009. Mitigating Flood Loss Through Local Comprehensive Planning in Florida (PhD thesis). Texas A&M University. Retrieved from: http://oaktrust.library.tamu.edu/bitstream/handle/1969.1/ETD-TAMU-2009-08-7161/KANG-DISSERTATION.pdf?sequence=3.

Kang, J.E., Peacock, W.G., Husein, R., 2010. An assessment of coastal zone hazard mitigation plans in Texas. J. Disaster Res. 5 (5), 520–528. https://doi.org/10.20965/jdr.2010.p0526.

Kim, H., Kakimoto, R., 2016. An international comparative analysis of local hazard mitigation plan evaluation for flood. Int. J. Disaster Resil. Built. Environ. 7 (4), 406–419. https://doi.org/10.1108/IJDRBE-07-2014-0056.

Kozel, D., 2015. Assessment of Avalanche Mitigation Planning for Developed Areas in the Rocky Mountain's of Colorado (Master thesis). University of Montana. Retrieved from: https://scholarworks.umt.edu/cgi/viewcontent.cgi?article=5561&context=etd.

Lavell, A., 2016. El programa Ciudades Sostenibles INDECI-PNUD, 2001 a 2016: Una evaluación. Retrieved from: https://erc.undp.org/evaluation/documents/download/9447.

Lyles, W., Stevens, M., 2014. Plan quality evaluation 1994–2012. J. Plan. Educ. Res. 34 (4), 433–450. https://doi.org/10.1177/0739456X14549752.

Lyles, W., Berke, P., Smith, G., 2014. A comparison of local hazard mitigation plan quality in six states, USA. Landsc. Urban Plan. 122, 89–99. https://doi.org/10.1016/j.landurbplan.2013.11.010.

Masterson, J.H., Peacock, W.G., Van Zandt, S.S., Grover, H., Schwarz, L.F., Cooper, J.T., 2014a. An assessment of hazard mitigation plans. In: Planning for Community Resilience: A Handbook for Reducing Vulnerability to Disasters. Island Press/Center for Resource Economics, Washington, DC, pp. 117–137, https://doi.org/10.5822/978-1-61091-586-1_7.

Masterson, J.H., Peacock, W.G., Van Zandt, S.S., Grover, H., Schwarz, L.F., Cooper, J.T., 2014b. Organizing and connecting through the disaster phases. In: Planning for Community Resilience: A Handbook for Reducing Vulnerability to Disasters. Island Press/Center for Resource Economics, Washington, DC, pp. 41–68, https://doi.org/10.5822/978-1-61091-586-1_3.

MEF, n.d.-a. Consulta del gasto del fondo para intervenciones ante la ocurrencia de desastres naturales-FONDES. Ministerio de Economía y Finanzas (MEF). Retrieved January 29, 2019, from: http://apps5.mineco.gob.pe/seguimiento_fondes/Navegador/default.aspx.

MEF, n.d.-b. Portal de Transparencia Económica, Consulta Amigable de Ejecución del gasto, Actualización Mensual. Ministerio de Economía y Finanzas (MEF). Retrieved July 1, 2019, from: http://apps5.mineco.gob.pe/transparencia/mensual/.

Meyerson, R., 2012. A Tool for Evaluating Plan Quality of Local Government Emergency Management Response Plans (Master thesis). University of North Carolina at Chapel Hill. Retrieved from: https://cdr.lib.unc.edu/indexablecontent/uuid:a1085eb7-c29b-493c-817e-48611f9d3a45.

Park, J., 2016. The Assessment of Quality of Comprehensive Plan for Storm and Flood Damage Reduction in Korea (Master thesis). Ulsan National Institute of Science and Technology. Retrieved from: https://scholarworks.unist.ac.kr/bitstream/201301/18352/1/000061.pdf.

PCM, 2017a. DS-N° 091-2017-PCM. Decreto Supremo que Aprueba el Plan de la Reconstrucción al que se Refiere la Ley N° 30556. Presidencia del Consejo de Ministros (PCM). Retrieved from: https://busquedas.elperuano.pe/download/url/decreto-supremo-que-aprueba-el-plan-de-la-reconstruccion-al-decreto-supremo-n-091-2017-pcm-1564235-1.

PCM, 2017b. Reconstrucción con Cambios. Respuesta, Rehabilitación y Reconstrucción. Diapositivas Presentadas por la PCM en el Congreso de la República el 24 de abril de 2017. Presidencia del Consejo de Ministros (PCM). Retrieved from: http://www.pcm.gob.pe/wp-content/uploads/2017/04/Reconstrucción-Con-Cambios.pdf.

PCM, 2018. Informe de Desempeño del PP 0068. Presidencia del Consejo de Ministros (PCM). Retrieved from: http://www.pcm.gob.pe/wp-content/uploads/2018/09/Informe-de-Desempeño-del-Programa-Presupuestal-0068_26.09.18.pdf.

Peacock, W.G., Kang, J.E., Husein, R., Burns, G.R., Prater, C., Brody, S., Kennedy, T., 2009. An Assessment of Coastal Zone Hazard Mitigation Plans in Texas. Texas A&M University Hazard Reduction and Recovery Center, College Statio, Texas, TX, Retrieved from: http://hrrc.arch.tamu.edu/_common/documents/09-01R_An_assessment_of_CZ_Haz_Mit_Plans_January_11,_2009.pdf.

RCC, 2017. Portafolio del plan integral de reconstrucción con cambios (modificado mediante D.S. N° 124-2017-PCM). Autoridad de la Reconstrucción con Cambios (RCC). Retrieved from: http://www.rcc.gob.pe/wp-content/uploads/2017/12/Portafolio_proyectos_Plan_Integral_Reconstrucción_Cambios_28DIC.xls.

Saunders, W., Grace, E., Beban, J., Johnston, D., 2015. Evaluating land use and emergency management plans for natural hazards as a function of good governance: a case study from New Zealand. Int. J. Disaster Risk Sci. 6 (1), 62–74. https://doi.org/10.1007/s13753-015-0039-4.

SENAMHI, 2017a. El Niño Costero 2017: Condiciones Termo-Pluviométricas a Nivel Nacional. Informe Técnico N° 028–2017/SENAMHI/DMA-SPC. Servicio Nacional de Meteorología e Hidrología del Perú (SENAMHI). Retrieved from: http://sigrid.cenepred.gob.pe/docs/PARAPUBLICAR/SENAMHI/Informe_Tecnico_N_28_2017_SEHAMHI_DMA_SPC_El_Nino-Costero_2017_Condiciones_Termo_Pluviometricas_a_Nivel_Nacional_Peru_2017.pdf.

SENAMHI, 2017b. Monitoreo de Lluvias en el Norte Enero—Febrero—marzo 2017. Servicio Nacional de Meteorología e Hidrología del Perú (SENAMHI). Retrieved from: http://www.senamhi.gob.pe/load/file/02231SENA-10.pdf.

Silapapiphat, A., 2015. The Review of Local Hazard Mitigation Plans in Ohio: What Local Factors Contribute Local Hazard Mitigation Plan Quality (PhD thesis). University of Akron. Retrieved from: https://etd.ohiolink.edu/!etd.send_file?accession=akron1430731923&disposition=inline.

Song, Y., Li, C., Olshansky, R., Zhang, Y., Xiao, Y., 2017. Are we planning for sustainable disaster recovery? Evaluating recovery plans after the Wenchuan earthquake. J. Environ. Plan. Manag. 60 (12), 2192–2216. https://doi.org/10.1080/09640568.2017.1282346.

Stevens, M.R., Shoubridge, J., 2015. Municipal hazard mitigation planning: a comparison of plans in British Columbia and the United States. J. Environ. Plan. Manag. 58 (11), 1988–2014. https://doi.org/10.1080/09640568.2014.973479.

UDEP, n.d. Mapa que reconstruye. Universidad de Piura (UDEP). Retrieved July 27, 2018, from: https://www.google.com/maps/d/viewer?usp=sharing_eil&mid=1lbOOwijy-u2ct_-HK4Xwhrm0Sd4.

UNISDR, 2015. Hacia el Desarrollo Sostenible: El Futuro de la Gestión del Riesgo de Desastres. Informe de Evaluación Global Sobre la Reducción del Riesgo de Desastres. United Nations Office for Disaster Risk Reduction (UNISDR). Retrieved from: https://www.preventionweb.net/english/hyogo/gar/2015/en/gar-pdf/GAR2015_SP.pdf.

UNISDR, 2016. United Nations General Assembly. Report of the Open-Ended Intergovernmental Expert Working Group on Indicators and Terminology Relating to Disaster Risk Reduction. United Nations Office for Disaster Risk Reduction (UNISDR). Retrieved from: https://www.preventionweb.net/files/50683_oiewgreportenglish.pdf.

Chapter 13

Business continuity as a means to strengthen disaster risk reduction in a coastal community of oyster farmers

Raymond S. Rodolfo[a,b] **and Mark R. Lapus**[b]
[a]*Ateneo de Manila University, Quezon City, Philippines,* [b]*Agriculture Sustainability Initiatives for Nature, Inc., Bulacan, Philippines*

1 Introduction

Oyster farming in the Philippines is an important livelihood in many coastal communities in the rural areas of the archipelago. These communities are usually composed of fishermen and informal settlers that are among the marginalized sectors of the country that live below the poverty line and are most vulnerable to natural hazards (Agbayani and Toledo, 2008; Islam et al., 2014; Suyo et al., 2013).

The Philippines sits in a region where climate and geophysical hazards such as tropical cyclones, tsunamis, earthquakes and volcanic eruptions are common and have caused significant losses of lives and damage to property (Lagmay et al., 2017). Due to climate variability, disasters will likely to be more common in coastal areas in the country (Orencio and Fujii, 2013).

According to the Hyogo Framework, disaster risk arises when hazards interact with physical, social, economic and environmental vulnerabilities (UNISDR, 2007). Thus, disaster risk reduction measures (DRR) are important in order to prevent loss of life and reduce damage to properties. This chapter looks at business continuity initiatives as a way to introduce DRR and enhance local inherent capacity and resilience to disasters.

2 Business continuity

Business continuity is defined as ensuring the continuation of key business activities after an adverse event such as natural disasters, with the human, material and financial resources available at the time. The reduction of resources is proportional to the magnitude and extent of the disaster and its direct and indirect impact on the enterprise and its suppliers, customers and clients along the same value chain. However, initiatives can be done in order to counteract the negative impacts of natural hazards on the continuity of business activities by strengthening resilience through risk reduction and mitigation measures such as disaster preparedness. Business continuity management initiatives is composed of three components: (1) Preventive measures; (2) Preparedness arrangements, both which are done prior to the occurrence of a disaster; and (3) Response actions when the disaster occurs (International Labour Office and ILO Programme for Crisis Response and Reconstruction, 2012).

3 Methods and study site

The study site is Sitio Ipil in Barangay Walay, Padre Burgos, Quezon Province, located in the southern part of Luzon Island in the Philippines. The community is composed of around 30 families, the majority of which live along the coast of an estuarine environment facing Tayabas Bay. The built up area of oyster farmers in the study site is centered at geographic coordinates 13°53.710′ N and 121°52.510′ E (Fig. 13.1). It was chosen because of the active partnership between the Agricultural Sustainability Initiatives for Nature (ASIN) Inc., a start-up company in the oyster industry which buys oysters and provide technical support to oyster farmers, and Ipil Action Group, a small community based organization of fishermen and oysters framers.

Our research aims to first develop a basic understanding of the current status of the oyster industry in the Philippines, including common growing methods, challenges and general profile of the oyster farmers. This is followed by a basic

Strengthening Disaster Risk Governance to Manage Disaster Risk. https://doi.org/10.1016/B978-0-12-818750-0.00013-1

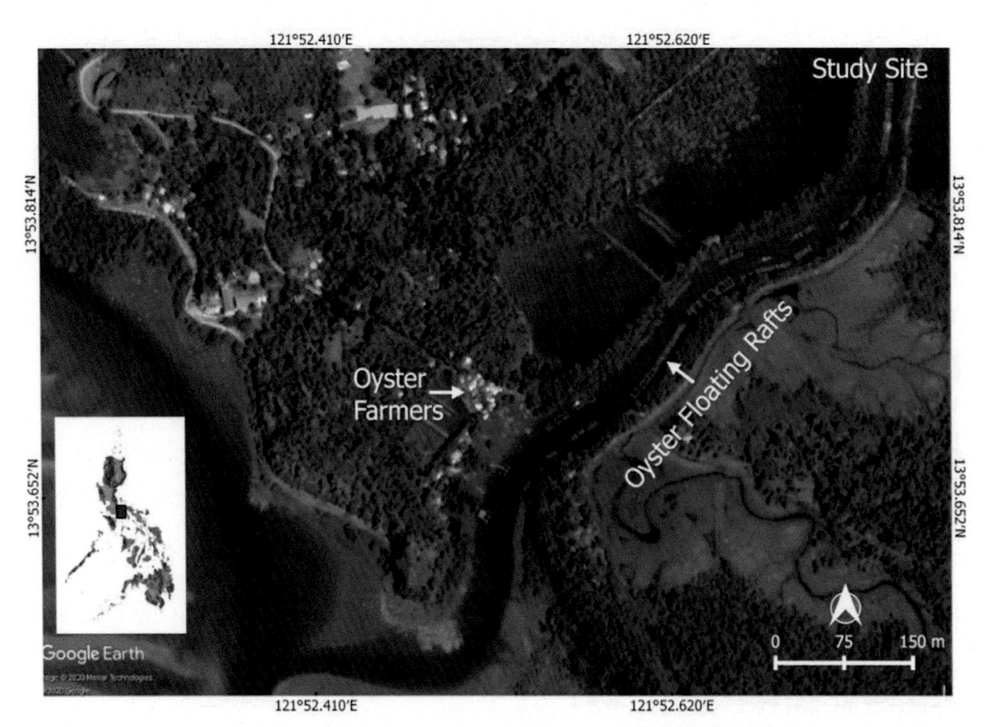

FIG. 13.1 Google Earth image of the study site showing the community of oyster farmers and their floating rafts.

geological hazard assessment of the study site using available maps and data sourced from government institutions such as the Philippine Astronomical and Geophysical Sciences Administration (PAGASA), Philippine Institute of Volcanology and Seismology (PHIVOLCS), Mines and Geosciences Bureau (MGB) and the University of the Philippines Nationwide Operational Assessment of Hazards (UP-NOAH).

Two semi-structured focus group discussions (FGD) were done from November 2017–2018 with the Ipil Action Group. The participants are characterized as a natural group belonging to one organization. Eight and ten participants were present in the two FGDs, with ages ranging from the mid-thirties to early sixties, the majority of which are men. To a certain extent, these participants are also considered as an expert group as they have been engaged in fishing most of their lives and are the incumbent or previous officers of their community based organization. The questions and discussions were aimed to draw out: (1) Their knowledge and understanding of natural hazards in their area and its effects on their livelihood and daily lives; and (2) Local gaps in disaster risk reduction and mitigation.

The potential impacts of these hazards to the oyster supply/value chain and business continuity were evaluated along with existing initiatives. Additional business continuity strategies were then suggested by drawing on the identified gaps and potentials for collaboration.

4 Results and discussion

4.1 Oyster farming in the Philippines

Oysters are filter feeder, bivalve mollusks that grow in brackish waters of estuaries which are common in the Philippines due to its archipelagic nature. The most common oyster species in the Philippines are the slipper-shaped oyster, *Crassostrea iredalei*; the palm-rooted oyster, *Saccostrea malabonensis*; *Saccostrea palmipes*; and the curly or wild oysters, *Saccostrea cucullata* (Panggat, 2016). Small-scale oyster farming is widespread in the Philippines due to: (1) The presence of coastal bays and rivers which are natural grounds for oysters; (2) The need for additional sources of income brought about by the dwindling catch of small-scale fishermen; and (3) Oysters have economic potential in both domestic and foreign markets (Samonte et al., 1994).

In the Philippines, the most common culture methods are broadcast, stake and rack. Table 13.1 lists the growing methods along with its advantages and disadvantages. Growers have specific areas to culture oysters within each respective municipalities. These areas are mostly family-run and highly generational unless otherwise, sold or leased to other fishermen in the area (Panggat, 2016). Stakes and spat collectors are installed in the natural oyster spawning grounds at the onset of the

TABLE 13.1 Summary of common oyster growing methods in the Philippines.

Method	Description	Advantages	Disadvantages
Broadcast method	Employed in shallow areas with firm solid bottoms to support the collectors such as empty oyster shells which are scattered over areas where there is natural setting occurs.	Simplest and low cost.	High mortalities of oysters due to siltation and predation. Difficulty in harvesting.
Stake method	Applied in relatively shallow areas with soft, muddy bottoms that would allow the bamboo poles to be driven into the substrate.	Higher growth rate and production per unit area.	Short life span of bamboo poles (1–2 years). Hinders the flow of water, causes siltation and shallowing of the water body.
Rack method	Uses bamboos or wooden frames to serve as substrates for oysters. Tires are often times used as cultches and spat collectors.	Higher growth rate and production per unit area.	Costly. Hinders the flow of water, causes siltation and shallowing of the water body.
Floating raft method	Refers to floating structures usually with hanging strings with cultches made from empty oyster shells, old rubber tires, sacks or coconut husks.	Higher growth rate and production per unit area. Can be towed to safety during typhoons.	Costly. Prone to poaching. May be washed downstream if not properly anchored.

rainy season, usually during the months of May to August. These months represent the peak period of oyster spawning in the Philippines when the environmental factors such as salinity and temperature are most favorable (Samonte et al., 1994). The drop in salinity in the estuary due to the influx of rain and flood water add stress to sexually mature oysters, inducing them to reproduce (Angell, 1986). Oysters are then allowed to grow to commercially viable sizes before they are harvested, which takes about 6–9 months (Panggat, 2016; Samonte et al., 1994).

Oyster farmers belong to the marginalized sector of society and often they do not own the land on which their houses are built. Oyster farmers also usually rely on their family members to labor for various activities such as the preparation of cultches, planting, harvesting, shucking and selling (Samonte et al., 1994). Additional problems encountered by oyster farmers include poaching, mortality due to siltation or sedimentation and pond effluents, absence of oyster spats, harmful algal blooms or red tide, lack of financing, lack of buyers and damage from river flooding and storm surge (Angell, 1986; Panggat, 2016; Samonte et al., 1994).

4.2 Geologic hazard assessment

The coastal community under study is situated at the mouth of Ipil River where the elevated portions are underlain by thinly bedded siltstone associated with slightly fossiliferous gray to brown shale, conglomerates and limy sandstone beds. The flat low lying coastal area is underlain by unconsolidated river and beach deposits mostly composed of clay, silt, sand and gravel (Bureau of Mines and Geo-Sciences, 1983). The climate in the immediate region is characterized as having little pronounced maximum rain period with a dry season that may last up to 3 months. In terms of historical typhoon frequency, the study site is in a region frequently visited by typhoons. Approximately 20 tropical cyclones enter the Philippine Area or Responsibility (PAR) each year, five of which are likely to be destructive (PAGASA, 2011).

Geologic hazards which include seismic, volcanic and hydro-meteorological hazards that may affect the coastal community were assessed for this study. The Philippines sits on a tectonically active area lined with faults and trenches. According to the Philippine Institute of Volcanology and Seismology, there have been approximately 90 destructive earthquakes and 40 tsunamis in the country in the past 400 years (Solidum, 2016). There are also 23 active volcanoes in the country stretching from north to south. The nearest active volcano is Mt. Banahaw, a stratovolcano complex, located 46 km northwest of the study site (PHIVOLCS, 2017b). Hydro-meteorological hazards include river flooding, storm surge or coastal flooding and rainfall induced landslides which may be triggered by intense and prolonged rainfall from weather systems in the country such as thunderstorms, typhoons, monsoon rains, tropical depressions, inter-tropical convergence zone and the tail end of the cold front (NDRRMC, 2013; NEDA-UNDP-ECHA, 2008).

Data and maps from government agencies and institutions were used to determine the susceptibility of the coastal community to geologic hazards. Fig. 13.2 shows a satellite image of the coastal community and the consequent landslide and storm surge hazard map available online at the UP-NOAH website (http://noah.up.edu.ph/#/).

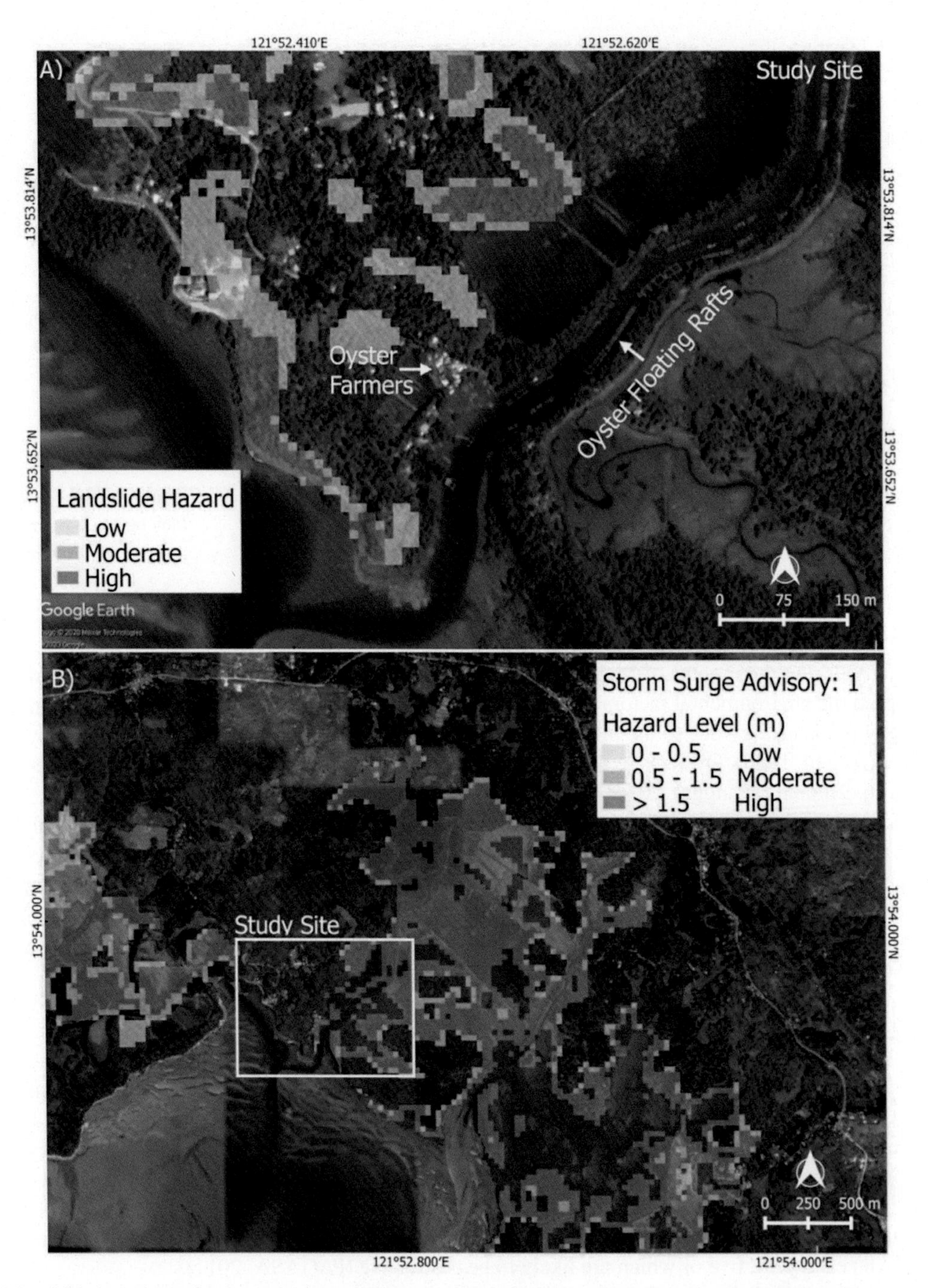

FIG. 13.2 (A) Landslide hazard map for the study site and (B) storm surge hazard map of the study site (storm surge of 2 m). *(Source: UP-NOAH overlain on Google Earth imagery.)*

In terms of seismic hazards, the study site is unlikely to be affected by ground rupture as the nearest identified active fault is 9.9 km away, according to the PHIVOLCS FaultFinder website, www.faultfinder.phivolcs.dost.gov.ph. The intensity of ground shaking the study site will experience is dependent on the location and magnitude of the earthquake. Based on the Philippine Earthquake Model, the study site is expected to experience a peak ground acceleration value of 0.4 g for an earthquake with 500-year return period (PHIVOLCS, 2017a). Areas underlain by quaternary alluvium are susceptible to liquefaction while steep and eroded slopes are susceptible to landslides during intense ground shaking. In terms of tsunamis, the study site can potentially be affected by locally generated tsunamis (PHIVOLCS, 2008). Volcanic hazards that may affect the coastal community is limited to ash fall especially if the prevailing winds are blowing south west from Mt. Banahaw.

Based on the landslide map of UP-NOAH, the steep slopes in the vicinity of the study site are susceptible to landslides and slope protection measures and continuous monitoring are recommended (Alejandrino et al., 2015). According to the flood hazard maps and anecdotal accounts, the coastal community is not affected by river flooding due to the small watershed of the Ipil River. However, it is susceptible to coastal flooding and storm surge. UP-NOAH identified storm surge vulnerable areas in the Philippines by simulating worst case scenarios over different coasts that are available at the UP-NOAH website (Lagmay et al., 2017; Lapidez et al., 2015). Fig. 13.2B shows the extent of inundation in the study site form a 2 m storm surge.

4.3 Business continuity initiatives and DRR

The susceptibility of the coastal community to geological hazards were assessed in order to determine how these hazards will affect the supply/value chain of oyster farming in the area. In general, these hazards may affect the oyster farmers either by damaging their houses or, at the worst, cause losses of lives. These hazards may also affect their livelihood by damaging their oyster farms and critical roads which lead to the market and urban centers. According to the United Nations Office for Disaster Risk Reduction (UNISDR), events of a hydro-meteorological origin constitute the large majority of disasters since they occur more frequently compared to seismic and volcanic hazards such as strong earthquakes and volcanic eruptions (UNISDR, 2007). This is alarming since among the geologic hazards that may affect the area, hydro-meteorological hazards, specifically storm surge, will have the greatest impact on the oyster livelihood in the study area. The common practice of growing oysters using the fixed stake and fixed rack methods are prone to damage during flood and storm surge events as they are stationary structures. They may also exacerbate the effects of flooding as these fixed structures impede water flow and cause siltation and the shallowing of the estuary.

The Bureau of Aquatic Resources of the Department of Agriculture (DA-BFAR) started distributing floating rafts to oyster farmers to discourage them from using the stake method. Floating rafts can be towed and safely anchored before an incoming typhoon (Fig. 13.3). However, a timely and accurate early warning system must be in place. Failure to warn oyster farmers of an incoming typhoon and storm surge may cause the floating rafts to be carried off shore by flood waters and may even destroy other aquaculture structures such as fish cages, similar to what happened in the study site during heavy rains in 2016.

In discussing DRR in the context of business continuity, several gaps were identified by the participants, which includes insufficient knowledge in the science of oyster farming and geologic hazards, lack of hazard early warning systems and community DRR plan, and concerns on environmental protection and management, and insufficient water, sanitation and hygiene (WASH) (Table 13.2). The emergence of environmental protection as a concern highlights the participants' understanding of how environmental quality will affect their livelihood of oyster farming and how environmental degradation may exacerbate the impact of disasters. They identified insufficient community water, sanitation and hygiene (WASH) initiatives as a threat to business continuity and to their personal and community's health and well-being. These concerns, however, are in resonance with the Sendai Framework as it broadens the scope of disaster risk reduction to cover environmental and biological hazards and risks and the promotion of health resilience (UNISDR, 2015).

The ongoing and proposed business continuity initiatives to address the identified gaps are anchored on a close collaboration between the community organization, business sector, government and higher educational institutions (HEIs) which hopefully will enable capacity development (Haigh et al., 2018). Optimistically, these initiatives will lead to resilience within the community which addresses geologic hazards, health, sanitation and the environment. It was highlighted during the focus group that all of these gaps need to be filled for the continuity of the oyster industry in the area and production and sustainability of high quality safe oysters.

In November 2018, a bulletin entitled "Impact Management of Weather Systems on Shellfish Aquaculture" was released by DA-BFAR and PAGASA. The bulletin was intended to explain the dynamics between shellfish farming with weather systems and its impact on day-to-day management and operations. It was the result of a participatory workshop among farmers from major shellfish farming areas, aquaculture scientists and weather/climate experts. Furthermore, DA-BFAR encourages its adoption in consideration of the local contextual knowledge application of the farmer-clients being served (Bureau of Fisheries and Aquatic Resources, 2018).

The bulletin covered the following weather systems that are active in the Philippines: Northeast and Southwest Monsoon, Easterlies, Intertropical Convergence Zone (ITCZ), Tail End of the Cold Front, Tropical Cyclone, High Pressure Area, Low Pressure Area and Localized Thunderstorm. It discussed the immediate impacts of these weather system on the farming habitat, operations and provided a generalized farm emergency response advisories against adverse weather/climatic conditions. This helped fill one of the gaps identified in this study which is the lack of a community DRR plan. It also affirms the importance of integrating business continuity planning in DRR. Results of the discussions between ASIN and the partner community were used to tailor fit the emergency response advisories based on their local context and existing collaborative activities (Table 13.3).

FIG. 13.3 *(Top)* Abandoned stake method showing siltation. *(Bottom)* Floating raft being distributed by the Bureau of Fisheries and Aquatic Resources and ASIN to the Ipil Action Group.

TABLE 13.2 Identified gaps and proposed measures and activities.

Identified gaps	Collaborative measures among the stakeholders and institutions
Insufficient knowledge of the science of oyster faming	Lectures and workshops to be conducted by technical experts from ASIN and government agencies.
Lack of a hazard early warning system	Establishment of a simple cellular network based early warning system for severe weather disturbances between ASIN and Ipil Action Group.
Insufficient knowledge on geologic hazards	Collaboration between ASIN, Ipil Action Group and Higher Education Institutions (HEIs) in conducting workshops on Geologic Hazards.
Lack of a community DRR plan	Collaboration between ASIN, Ipil Action Group, Local Government Unit (LGU), government agencies and HEIs in crafting a community based DRR plan.
Environmental protection and management	Collaboration between ASIN, Ipil Action Group, LGU, government agencies and HEIs in conducting lectures and workshops. Recommend, develop and strengthen local policies and laws to safeguard environment.
Insufficient water, sanitation and hygiene (WASH)	Collaboration between ASIN, Ipil Action Group, LGU, government agencies and HEIs in promoting and sustaining WASH initiatives.

TABLE 13.3 Integrated Disaster Risk Reduction—Business Continuity Strategies for a Tropical Typhoon adapted from Impact Management of Weather Systems on Shellfish Aquaculture (Bureau of Fisheries and Aquatic Resources, 2018).

Key operations response (preventive measures) of the community	ASIN business continuity measures
Low pressure area develops	
Monitor radio and television news from PAGASA. Check the integrity of floating rafts for priority/early reinforcement. Acquire the necessary materials and supplies for reinforcement.	Monitor weather updates from PAGASA and relay to the community.
Tropical depression/storm/typhoon enters the Philippine area of responsibility (PAR)	
Continue to monitor weather updates while conducting detailed inspection of rafts to identify priority areas for quick reinforcement needs and mobilize team for emergency repairs. Conduct quick stock assessment and determine availability of marketable oysters and inform ASIN regarding estimated volume. Prepare to conduct emergency partial-harvest. Selected members should conduct logistics for emergency evacuation and prepare the necessary supplies.	Update the community on the status of the weather disturbance. Prepare to visit the community to purchase the partial-harvest. Prepare contingency funds for the repairs and evacuation needs.
Immediately after typhoon alert warning raised over the farming area	
Continue with repairs and monitor weather updates. Prepare to transfer the floating rafts to a safe and secure area away from strong waves and currents. Start the partial-harvesting of marketable oysters to ensure partial recovery of investment expense. Continue contingency planning for an emergency evacuation.	Update the community on the status of the weather disturbance. Visit the community to purchase and pick-up the partial-harvest.
72 h before impact	
Monitor weather updates and follow PAGASA advisories. Complete the reinforcement of floating rafts and transfer to a safe area. Continue the partial-harvesting of marketable oysters to ensure partial recovery of investment expense. Prepare an evacuation plan and necessary emergency supplies.	Update the community on the status of the weather disturbance. Visit the community to purchase and pick-up the partial-harvest.
36–48 h before impact	
Monitor weather updates and follow PAGASA advisories. Complete the partial-harvesting of marketable oysters. Be ready to execute the evacuation plan.	Update the community on the status of the weather disturbance. Visit the community to purchase and pick-up the emergency partial-harvest.
18–24 h before impact	
Monitor weather updates and follow PAGASA advisories. Continue final emergency harvesting if conditions permit. Start planning for relief, rehabilitation and business continuity measures.	Update the community on the status of the weather disturbance. Visit the community to purchase the partial-harvest.
12 h before impact	
Monitor weather updates and follow PAGASA advisories. Conduct final evacuation if necessary.	Update the community on the status of the weather disturbance.
During impact (typhoon affecting the farm site)	
Monitor weather updates and follow PAGASA advisories. Consult the local Disaster Risk Reduction and Management Office (DRRMO) for further advice and support. Draw a concrete plan for emergency rehabilitation.	Update the community on the status of the weather disturbance.
Immediately after impact	
Monitor weather updates and follow PAGASA advisories. Consult the local DRRMO for further advice and support. Submit damage reports to BFAR, ASIN and local DRRMO. Prioritize the retrieval/recovery of submerged floating rafts and conduct a quick assessment for repairs. Conduct priority repairs to damaged houses, critical structures and rafts.	Aid in the rehabilitation efforts and repairs to damaged rafts. Distribute oyster cultches with attached spats from the hatchery to replace the harvested oysters.

The UNISDR presents "Five Essentials for Business in Disaster Risk Reduction," which are: (1) To promote public-private partnerships for resilience; (2) To leverage private sector expertise and strengths; (3) To foster a collaborative exchange and dissemination of data; (4) To support national and local risk assessment; (5) To support the development and strengthening of national and local laws, regulations, policies and programs (UNISDR, 2013). Over time, these essential components are being applied to the partnership between ASIN and the Ipil Action Group with the support of DA-BFAR, local government units and other institutions. ASIN is in the process of continuing the DA-BFAR initiative of distributing floating rafts to the Ipil Action Group in order to promote a more disaster resilient oyster growing method which is essential in ensuring the continuous supply of oysters. The success of this method can only be made possible when combined with a timely and accurate early warning system, a simple and open communication system where ASIN could continuously update and provide advice for the coastal community about weather hazards (Bureau of Fisheries and Aquatic Resources, 2018). This early warning system and communication protocol will also apply for tsunami warnings and other instances where critical information needs to be provided to the community as soon as possible. The activity to ensure the safety of oyster rafts and oyster farmers' livelihood during typhoons will help strengthen DRR in the coastal community.

5 Conclusions

This study highlights the importance of collaboration between the community members, business sector and government agencies. Higher education institutions were also identified to perform an important role in terms of hazard research and information dissemination. The findings of this study also shows that DRR initiatives can be integrated with business continuity management components such as: (1) Preventive measures which in this case involves the use of floating rafts which is a more sustainable and flood resilient oyster growing method; (2) Preparedness arrangements which includes hazard assessment, information dissemination and establishment of protocols and strategies which will be followed prior to extreme natural events; and (3) Response actions in terms of actual business continuity measures and action points.

By looking at community oyster farming in the Philippines, business continuity as well as DRR initiatives are hand in hand in increasing the resilience of these communities to disasters by enabling them with proper knowledge and continuous support. Sound and sustainable business practices can provide stable, long-term livelihood opportunities which will also be able to reduce the risk of the oyster farmers to disasters and transform them into a resilient community by helping them overcome poverty.

Acknowledgments

This study was inspired by the need for higher educational institutions to work for resilience education in coastal communities, raised during the Erasmus + CApacity Building in Asia for Resilience EducaTion (CABARET) workshop. The European Commission support for the production of this publication does not constitute an endorsement of the contents which reflects the views only of the authors, and the Commission cannot be held responsible for any use which may be made of the information contained therein.

References

Agbayani, R.F., Toledo, J.D., 2008. Institutional capacity development for sustainable aquaculture and fisheries: strategic partnership with local institutions. In: Fisheries for Global Welfare and Environment Memorial Book of the 5th World Fisheries Congress 2008, pp. 435–448. http://www.terrapub.co.jp/onlineproceedings/fs/wfc2008/index.html.

Alejandrino, I., Aquino-Chow, D., Ariola, H., Bonus, A., Eco, R., Escape, C., Felix, R., Ferrer, P., Gacusan, R., Galang, J., Herrero, T., Llanes, F., Luzon, P., Montalbo, K., Obrique, J., Ortiz, I., Quina, C., Rabonza, M., Realino, V., Sabado, J., Sulapas, J., 2015. In: Lagmay, A.M.F. (Ed.), Landslide Hazard Map Atlas: Quezon. University of the Philippines Press, Manila.

Angell, C.L., 1986. The Biology and Culture of Tropical Oysters. ICLARM. http://pubs.iclarm.net/libinfo/Pdf/Pub%20SR76%2013.pdf.

Bureau of Fisheries and Aquatic Resources, 2018. Impact Management of Weather Systems on Shellfish Aquaculture. BFAR Aquaculture Technology Bulletin, vol. 43.

Bureau of Mines and Geo-Sciences, 1983. Geological Map of Unisan Quadrangle. Sheet 3361-I, first ed. (Map).

Haigh, R., Amaratunga, D., Hemachandra, K., 2018. A capacity analysis framework for multi-hazard early warning in coastal communities. Procedia Eng. 212, 1139–1146. https://doi.org/10.1016/j.proeng.2018.01.147.

International Labour Office, & ILO Programme for Crisis Response and Reconstruction, 2012. Multi-Hazard Business Continuity Management: Guide for Small and Medium Enterprises. ILO, Geneva.

Islam, M.M., Sallu, S., Hubacek, K., Paavola, J., 2014. Vulnerability of fishery-based livelihoods to the impacts of climate variability and change: insights from coastal Bangladesh. Reg. Environ. Chang. 14 (1), 281–294. https://doi.org/10.1007/s10113-013-0487-6.

Lagmay, A.M.F., Racoma, B.A., Aracan, K.A., Alconis-Ayco, J., Saddi, I.L., 2017. Disseminating near-real-time hazards information and flood maps in the Philippines through Web-GIS. J. Environ. Sci. 59, 13–23. https://doi.org/10.1016/j.jes.2017.03.014.

Lapidez, J.P., Tablazon, J., Dasallas, L., Gonzalo, L.A., Cabacaba, K.M., Ramos, M.M.A., Suarez, J.K., Santiago, J., Lagmay, A.M.F., Malano, V., 2015. Identification of storm surge vulnerable areas in the Philippines through the simulation of Typhoon Haiyan-induced storm surge levels over historical storm tracks. Nat. Hazards Earth Syst. Sci. 15 (7), 1473–1481. https://doi.org/10.5194/nhess-15-1473-2015.

NDRRMC, 2013. The Broadcaster's InfoChart on Emergency Preparedness. https://www.phivolcs.dost.gov.ph/images/Manual-the-broadcasters-infochart-2014.pdf.

NEDA-UNDP-ECHA, 2008. Mainstreaming Disaster Risk Reduction in Subnational Development Land Use/Physical Planning in the Philippines. National Economic and Development Authority, United Nations Development Programme and European Commission Humanitarian Aid. http://www.ph.undp.org/content/philippines/en/home/library/environment_energy/guideline-mainstreaming-drr.html.

Orencio, P.M., Fujii, M., 2013. A localized disaster-resilience index to assess coastal communities based on an analytic hierarchy process (AHP). Int. J. Disaster Risk Reduct. 3, 62–75. https://doi.org/10.1016/j.ijdrr.2012.11.006.

PAGASA, 2011. Climate Change in the Philippines. PAGASA Climatology and Agrometeorology Division. http://dilg.gov.ph/PDF_File/reports_resources/DILG-Resources-2012130-2ef223f591.pdf.

Panggat, E.B., 2016. Industry Roadmap: Processed Bivalve Molluscs (EU-Philippines Trade Related Technical Assistance Project 3). AECOM.

PHIVOLCS, 2008, October 30. Tsunami Prone Areas in the Philippines. http://www.phivolcs.dost.gov.ph/index.php?option=com_content&view=article&id=312&Itemid=500027.

PHIVOLCS, 2017a. The Philippine Earthquake Model: A Probabilistic Seismic Hazard Assessment of the Philippines and Metro Manila. Philippine Institute of Volcanology and Seismology.

PHIVOLCS, 2017b, December 14. Active Volcanoes. http://www.phivolcs.dost.gov.ph/index.php?option=com_content&view=article&id=8235%3Aactive-volcanoes&catid=55&Itemid=86.

Samonte, G.P.B., Siar, S.V., Ortega, R.S., Espada, L.T., 1994. Socio-Economics of Oyster and Mussel Farming in Western Visayas, Philippines. pp. 1069–1078. http://repository.seafdec.org.ph/handle/10862/264.

Solidum, R.U., 2016, January 12. Tsunami Disaster Management in the Philippines. https://www.pari.go.jp/special/special3/files/items/common/File/160112symp-14solidum.pdf.

Suyo, J.G., Prieto-Carolino, A., Subade, R., 2013. Reducing and managing disaster risk through coastal resource management: a Philippine case. Asian Fish. Sci. 26, 198–211.

UNISDR (United Nations Office for Disaster Risk Reduction), 2007. Hyogo Framework for Action 2005–2015: Building the Resilience of Nations and Communities to Disasters. United Nations Office for Disaster Risk Reduction. https://www.preventionweb.net/publications/view/1037.

UNISDR (United Nations Office for Disaster Risk Reduction), 2013. Business and Disaster Risk Reduction, Good Practices and Case Studies. UN Office for Disaster Risk Reduction (UNISDR). https://www.unisdr.org/files/33428_334285essentialscasestudies.pdf.

UNISDR (United Nations Office for Disaster Risk Reduction), 2015. Sendai Framework for Disaster Risk Reduction 2015-2030. https://www.unisdr.org/we/coordinate/sendai-framework.

Index

Note: Page numbers followed by *f* indicate figures, *t* indicate tables, and *b* indicate boxes.

S

T

U

V

W

X

Y

Z

Printed in the United States
By Bookmasters